YOUR KNOWLEDGE HAS VALUE

D1827232

- We will publish your bachelor's and
  master's thesis, essays and papers

- Your own eBook and book -
  sold worldwide in all relevant shops

- Earn money with each sale

Upload your text at www.GRIN.com
and publish for free

Bibliographic information published by the German National Library:

The German National Library lists this publication in the National Bibliography; detailed bibliographic data are available on the Internet at http://dnb.dnb.de .

Imprint:

Copyright © 2015 GRIN Verlag, Open Publishing GmbH
Print and binding: Books on Demand GmbH, Norderstedt Germany
ISBN: 9783668352452

This book at GRIN:

http://www.grin.com/en/e-book/344961/optimization-of-dna-concentration-in-rapd-fingerprinting-of-phytophthora

Nesaratnam Alwar

Optimization of DNA concentration in RAPD fingerprinting of Phytophthora infestans

GRIN Publishing

GRIN - Your knowledge has value

Since its foundation in 1998, GRIN has specialized in publishing academic texts by students, college teachers and other academics as e-book and printed book. The website www.grin.com is an ideal platform for presenting term papers, final papers, scientific essays, dissertations and specialist books.

Visit us on the internet:

http://www.grin.com/

http://www.facebook.com/grincom

http://www.twitter.com/grin_com

Optimization of DNA concentration in RAPD fingerprinting of *Phytophthora infestans*

Submitted by:

ALWAR Nesaratnam Nagan

Project submitted in partial fulfillment of the
requirement of
BSc (Hons) Biology

University of Mauritius

Faculty of Science

Department of Biosciences

March 2015

TABLE OF CONTENTS

Chapter 3 Methodology

Chapter 4 Results

Chapter 5 Discussion

LIST OF TABLES

LIST OF FIGURES

ACKNOWLEDGEMENTS

A completed project work bears only the name of the student; however the path leading to its completion is always achieved in combination with the dedicated action of other people. I hence wish to acknowledge my appreciation to certain people that have been of great help to me during these past several months.

First, I would like to thank my parents and sister for their unconditional contributions and support.

I would also like to thank Dr N. Taleb-Hossenkhan, for her advices, continued guidance and assistance.

I am grateful to Keshav Gangadin from the Food and Agricultural Research and Extension Institute (FAREI) for his expert guidance on field.

Thanks also to Mrs. Anishta for her valuable help in the Molecular Laboratory.

DISSERTATION DECLARATION FORM

UNIVERSITY OF MAURITIUS

PROJECT/DISSERTATION DECLARATION FORM

Name:	
Student ID:	
Programme of Studies:	
Module Code/Name:	
Title of Project/Dissertation:	
Name of Supervisor(s):	

Declaration:

In accordance with the appropriate regulations, I hereby submit the above dissertation for examination and I declare that:

(i)   I have read and understood the sections on **Plagiarism and Fabrication and Falsification of Results** found in the University's "General Information to Students" Handbook (20..../20....) and certify that the dissertation embodies the results of my own work.

(ii) I have no objection to submit a soft copy of my dissertation through the Turnitin Platform. I confirm that the hard copies and soft copies, including the one uploaded through the Turnitin Platform, in the final assignment submission link indicated by the Programme/Project Coordinator, are identical in content.

(iii) I have adhered to the 'Harvard system of referencing' or a system acceptable as per "The University of Mauritius Referencing Guide" for referencing, quotations and citations in my dissertation. Each contribution to, and quotation in my dissertation from the work of other people has been attributed, and has been cited and referenced.

(iv) I have not allowed and will not allow anyone to copy my work with the intention of passing it off as his or her own work.

(v)   I am aware that I may have to forfeit the certificate/diploma/degree in the event that plagiarism has been detected after the award.

(vi) Notwithstanding the supervision provided to me by the University of Mauritius, I warrant that any alleged act(s) of plagiarism during my stay as registered student of the University of Mauritius is entirely my own responsibility and the University of Mauritius and/or its employees shall under no circumstances whatsoever be under any liability of any kind in respect of the aforesaid act(s) of plagiarism.

Signature:	Date:

ABSTRACT

Phytophthora infestans is a pathogenic oomycete which causes the late blight disease affecting both potato and tomato plantations The *P.infestans* populations in Mauritius have not yet been genetically characterized to assess the possible strains present on the island. Random Amplified Polymorphic DNA (RAPD) is a low cost and simple genetic characterization tool that can be used to genetically characterize the different strains of *P.infestans* and lead towards a better management of the late blight disease. However, the RAPD fingerprinting is one which requires an extensive optimization in terms of the conditions and the adherence to a stringent protocol.

The aim of this study was to design and apply a series of experiments to optimize the RAPD protocol through the use of a set of DNA template concentrations. In this study, genomic DNA was extracted from 2 *P.infestans* isolates originating from potato and 1 *P.infestans* isolate emanating from tomato. The genomic DNA obtained from each isolates was diluted to obtain a set of DNA concentrations which were used for the screening of 30 RAPD primers and for further testing to identify the best DNA template concentration. The clarity of the amplified DNA fragments obtained during electrophoresis was used to determine the optimal DNA template concentration in this study.

The RAPD primers were screened at DNA template concentrations 20, 50, 80, 100 and 200ng/µl. 6 RAPD primers were selected and tested with DNA template concentrations of 30, 50 and 70ng/µl. DNA template concentrations of 30 and 50ng/µl gave consistent results with regard to clarity of amplified DNA and these were compared with a DNA template concentration of 40ng/µl. In this study, it was found that a DNA template concentration of 40ng/µl gave the best result in terms of clarity of the amplified DNA fragments and the reproducibility of RAPD ranged between 30 and 50ng/µl.

The RAPD protocol requires the optimization of all the parameters before valid claims can be made about the level of genetic diversity of *P.infestans* populations. Since RAPD is low cost and simple to undertake, it can be a useful tool to assess the diversity of strains of *P.infestans* causing late blight infections annually.

Keywords: Genetic characterization, late blight disease, optimization, *Phytophthora infestans*, Random Amplified Polymorphic DNA (RAPD)

LIST OF ABBREVIATIONS

ABBREVIATION	DESCRIPTION
A_{260}	Absorption at 260nm
A_{280}	Absorption at 280nm
ß-ME	Beta-Mercaptoethanol
bp	base pair
cm	centimetre
°C	degree Celsius
EDTA	Ethylenediaminetetraacetic acid
g	gram
M	molar
mg	milligram
ml	millilitre
mM	millimolar
µl	microlitre
µM	micromolar
NaCl	Sodium chloride
PCR	Polymerase chain reaction
RAPD	Random Amplification of Polymorphic DNA
RFLP	Restriction Fragment Length Polymorphism
rpm	revolutions per minute
SDS	Sodium dodecyl sulfate

Chapter 1
Introduction

1.1 Introduction

Phytophthora infestans which originates from the mountainous regions of Central Mexico (Goss *et al.*, 2014), causes the late blight disease. During the 1840s, late blight caused extensive damage in agricultural fields and food shortages throughout Europe. The severe impact of the disease at that time in Ireland had caused the "Irish Potato Famine". Late blight disease can eradicate an entire field of potatoes in a few days under suitable weather conditions (Burges *et al.*, 2005). *Phytophthora infestans* is considered to be an important pathogen of potato and tomato production systems worldwide (Grünwald & Flier, 2005) and a high chemical input is required in order to control the disease (Céspedes *et al.*, 2012). It is mainly due to the fact that the pathogenic oomycete has been able to evolve and overcome a great majority of the control measures that have been introduced over the years. Late blight has consequently established itself as one of the major limiting factors affecting potato production in the world in recent years. Outbreaks of late blight are regularly reported each year in Mauritius in both potato and tomato growing areas.

In order to develop effective strategies for the management of late blight, it is vital to understand the epidemiology of the disease and to closely monitor the different strains present in the country. Identification of the oomycete under the microscope is possible through the presence of lemon shaped sporangia but it is impossible to distinguish between the different strains because they all look the same. Therefore, the differentiation of the strains relies uniquely on DNA-based sequence approaches (Martin *et al.*, 2012). Genetically characterizing the local strains can therefore provide valuable information such as whether it is the same strains that are causing infection each year; the number of strains present on the island; how the different potato cultivars respond to these strains and to what extent those strains exhibit host specificity.

Rapid identification of the strains is important as new genotypes which are more resistant to systemic fungicides can appear. A variety of molecular markers are available to genetically characterize strains of *Phytophthora infestans* and one of the simplest methods now commonly used for genetic diversity studies is Random Amplification of Polymorphic DNA (RAPD). The speed, simplicity, low cost and quality of the RAPD technique have made it widely popular as a genetic characterization tool (Kumar & Gurusubramanian, 2011). RAPD technique enables the generation of a considerable amount of genetic markers by using small amounts of DNA and there is no need for any other form of molecular characterization, such as cloning or sequencing of the concerned genome (Bardakci, 2001). However, the use of RAPD-PCR as a genetic characterization tool requires an extensive optimization in terms of template DNA concentration, primer concentration and magnesium chloride concentration.

1.2 Aim

The aim of this project is to design and carry out a set of experiments in order to optimize the utilization of RAPD-PCR for the genetic characterization of local populations of *P.infestans*, with respect to the starting concentration of template DNA.

1.3 Objectives

The specific objectives of the study involve:

- Collection of potato leaves infected with *P.infestans* strains from the field.
- Culture of the different *P.infestans* strains obtained from infected leaves and available isolates on Rye B medium.
- Performing DNA extraction from the cultured strains.
- Screening of 30 RAPD primers to identify which ones provide the best interpretable results and these will be selected for further testing on a range of DNA concentrations.
- Finding the optimum DNA concentration for RAPD-PCR through testing of various DNA template concentrations.

Chapter 2
Literature Review

2.1 Biology and Taxonomy of *Phytophthora infestans*

2.1.1 Biology of *Phytophthora infestans*

Phytophthora infestans was named by Anton de Bary, who had also elucidated the life cycle of this fungus in 1876. Formerly, the fungus was named *Botrytis infestans* in the mid-19th century by Jean Montagne, however, de Bary discovered that the late blight fungus did not share the same characteristics as those species in the *Botrytis* genus and he thus created the *Phytophthora* genus. The name *Phytophthora* is generated from the combination of Greek words: '*Phyto*' meaning plant and '*phthora*' meaning destroyer (Schumann & D'Arcy, 2000).

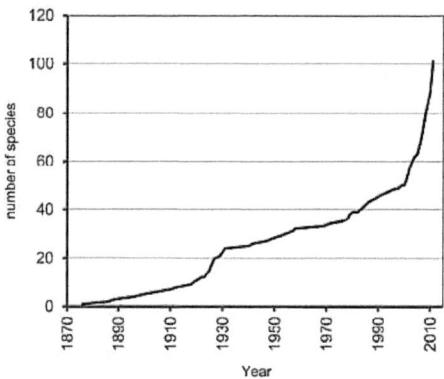

Figure 2.1: Increase in the number of described *Phytophthora* species over time (Kroon, 2012).

By the year 2010, as shown in the graph above, 101 *Phytophthora* species had been successfully described and recognized (Kroon, 2012). Of all these species, *Phytophthora infestans* is the most famous one, due to its damaging effects on potato and tomato cultivations worldwide. Also, unlike other species of *Phytophthora* which cause rotting mainly at the level of the roots, late blight infections affect the leaves, stems, potato tubers and tomato fruits.

Phytophthora infestans is a eukaryote as it contains nuclei and other membrane bound organelles. It produces microscopic, hyaline and lemon shaped asexual spores which are known as sporangia (Shihab & Ahmad, 2014). These sporangia are thin-walled and have a comparatively short lifespan outside living host tissue. *P.infestans* is assigned to the kingdom Stramenopila because it produces zoospores that have two hollow hair-like flagella and which is a particular feature of that kingdom. A whiplash flagellum is located on the posterior end of the zoospore to allow forward movement and a tinsel flagellum which is fibrous and ciliated, is found on the anterior end to pull the spore through water (Volk, 2001). *Phytophthora infestans*

3

is grouped together with some plant pathogens that are commonly known as "water molds", as they have a high affinity for water (Schumann & D'arcy, 2000).

For a long period of time in the past, *Phytophthora infestans* had traditionally been considered as a member of the kingdom Fungi because of its biological, ecological and epidemiological characteristics. For example, they produce hyphae, they acquire their nutrients by absorption and they produce filamentous threads known as mycelium which is a characteristic of true fungi. Nonetheless, present-day molecular and biochemical investigations advocate that oomycetes possess a very limited taxonomic relationship with the true fungi but they instead fit in the kingdom Stramenopila which is one of several eukaryotic kingdoms (Kamoun, 2003). Furthermore, the size of *P.infestans* genome is estimated to be about 240Mb whereas the genome size of the true fungi is about 40Mb (Haas *et al.*, 2009; Kupfer *et al.*, 1997).

2.1.2   Taxonomy of *Phytophthora infestans*

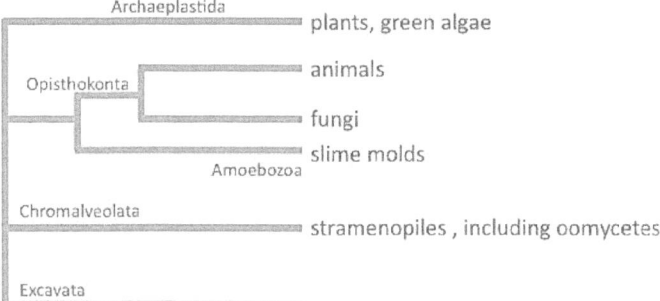

Figure 2.2: The relationship between the different supergroups and their kingdoms (Carris *et al.*, 2012).

In the above phylogenetic classification, it can be seen that the true fungi share a closer evolutionary relationship with the kingdom Animalia in the supergroup Opisthokonta. The oomycetes on the other hand, are classified in the supergroup Chromalveolata and in the kingdom Stramenopila. rRNA based studies confirm the fact that the oomycetes share a closer phylogenetic relationship with the diatoms and brown algae (Sogin & Silberman, 1998).

In the fourth edition of 'Introductory Mycology', Alexopoulos *et al.* (1996, cited in Soni & Soni, 2010, p.103) assigned the genus *Phytophthora* to the:

- Kingdom Stramenopila,
- Phylum Oomycota,
- Class Oomycetes,
- Order Perenosporales,
- Family Pythiaceae

The family Pythiaceae is further subdivided into two genera, *Phytophthora* and *Pythium*. The genus *Phytophthora* contains many species including *P.infestans* that cause damage to commercially important crops like potato, tomato, strawberries and papaya.

2.2 Origin and migration routes of *Phytophthora infestans*

The identification of the centre of origin of *Phytophthora infestans* was subjected to vigorous debates in the 1990s when two different established theories on the origin of the first inoculum collided: the Andean theory and the Mexican theory. Analyses of microsatellite markers and sequences of four nuclear genes provided information that suggests that *P.infestans* is of Mexican origin (Goss *et al.*, 2014). The center of genetic diversity of *P.infestans* in the Toluca Valley, Mexico, along with the certitude that it undergoes sexual reproduction and that both mating types occur in equal frequencies in that region also favour the Mexican theory (Grünwald and Flier, 2005). The findings support the idea that the Andean population is a descendant of the populations in the Central Valley of Mexico (Goss *et al.*, 2014).

The A1 and A2 mating types of *P.infestans* were both identified in Mexico, however, populations that had been sampled outside Mexico were made up of the A1 type and were thus asexual (Shaw & Wattier, 2003, p. 23). Investigations at the level of nuclear DNA markers and mitochondrial haplotype of the oomycete by researchers at Cornell University in the 1990s revealed a common lineage of *P.infestans* named US-1 which occurred in 13 different countries across four continents (Goodwin *et al.*, 1994). The A1 mating type which is widespread in the United States of America, Canada and the European region is associated with the US-1 genotype. Many plant pathologists had hypothesized that the migration of *P.infestans* occurred from the Mexican land to the United States of America and then to Europe in the 1840s and had caused severe crop damage on a global scale (Goodwin *et al.*, 1994). However, the results obtained from analyses of strains that had been sampled during the period of the epidemic in the 1840s exposed a different genotype from the US-1 lineage (Ristaino, 2002).

The A2 mating type was observed in the late 1970s in parts of Europe (Fry *et al.*, 1993). The dissemination of *Phytophthora* species occurred through the maritime transportation of potatoes from Mexico towards Europe. The migration of both the A1 and A2 strains from Mexico and their asexual or sexual mating induced greater variations in the populations of *P.infestans*. The confirmation of new strains having increased aggressiveness, the ability to resist to fungicides and different responses to environmental parameters has accumulated (Fry and Smart, 1999).

In the early 1980s, after the discovery of the A2 mating type, identification of 'new' populations of A1 mating type were identified using genetic markers. The 'old' US-1 population was displaced by the 'new' mating type populations by an incredible speed indicating that the 'new' population had a fitness advantage as compared to the 'old' one (Guo *et al.*, 2009)

2.2.1 *Phytophthora infestans* in Mauritius

Several strains of the pathogen are known to exist and the ones present in Mauritius have not yet been fully characterized. Molecular characterization of *Phytophthora infestans* at Cornell University, USA, showed that isolates from *Solanum tuberosum* and *S. lycopersicum* received from Mauritius belonged to the US-1 genotype, the most widely distributed strain (Mauritius Sugar Industry Research Institute, Annual Report, 2000, p.48).

2.3   The disease cycle of *Phytophthora infestans*

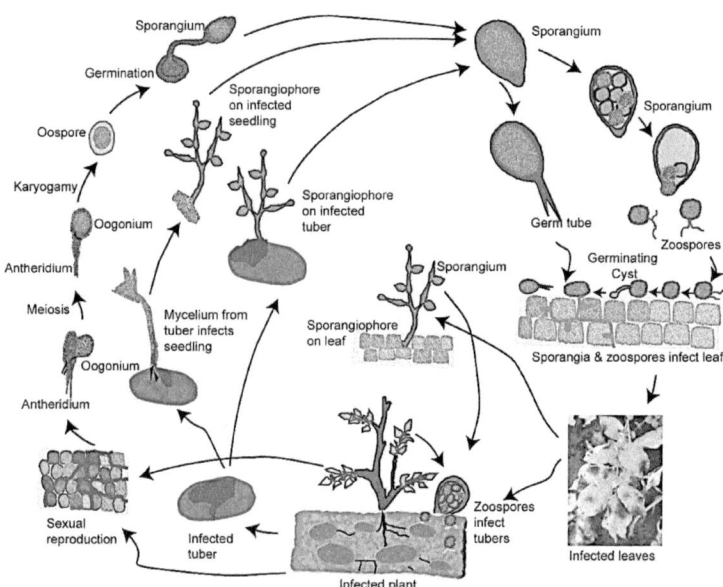

Figure 2.3: Life cycle of *Phytophthora infestans* (Agrios, 2005, p.425).

P. infestans can undergo both asexual and sexual reproduction as shown in Figure 2.3. The asexual cycle enables a rapid growth of the population. The damages caused by the late blight disease are due to the ability of the oomycete to produce asexual spores that are airborne or waterborne and that propagate from one host to another in the environment. The propagation of late blight epidemics relies on humidity and temperature of the environment, whereby a high humidity and temperature range between 15-25°C will favour the growth of the oomycete whereas at higher temperatures the growth is halted (Agrios, 2005, p.421).

7

2.3.1   Asexual reproduction of *Phytophthora infestans*

The fungus reproduces asexually when it has absorbed sufficient nutrients and if the atmospheric conditions are suitable. Late blight infections are initiated from the sporangia which release zoospores or produce germ tubes. Dispersal of the sporangia occurs over long distances by wind (Aylor *et al.*, 2001) or on short distances when the sporangia get splashed by rain and land on the surface of potato leaves. These biflagellated zoospores swim for a while until they locate the leaf stomata. They eventually attach themselves; encyst, whereby the zoospores lose their flagellum and form germ tubes on the leaf surface. The penetration into the leaf occurs when the apex of the germ tube forms an appressorium which enables the invasion of the underlying host cells (Birch & Whisson, 2001). This type of asexual reproduction is known as indirect germination (MetaPathogen, no date). When the temperature is above 15°C the sporangia undergoes direct germination to form the germ tubes (MetaPathogen, no date). Within the leaf, growth of the hyphae produces haustoria; structures that are specialised in retrieving nutrients from the host plant (MetaPathogen, no date). The fungus will establish a biotrophic interaction up to 48 hours without causing any visible lesions and is followed by necrotrophic interaction. Lesions are thus formed on the leaves. The mycelia will grow through the stomata after 3 to 5 days and produce new sporangia (Kamoun *et al.*, 1998). About 100,000 sporangia can be produced by a single lesion within a day and these sporangia may reach the tubers or neighbouring leaves of other plants and trigger infection during wet and cool conditions.

2.3.2   Sexual reproduction of *Phytophthora infestans*

Phytophthora infestans has the ability to reproduce both sexually and asexually. The pathogen was thought to be asexual until the 1950s (Shaw and Wattier, 2003, p.23). This oomycete is heterothallic and it consists of the A1 and A2 mating types. When specific hormones are produced from the mating types, sexual differentiation occurs and the oogonia and the antheridia are formed within a mating zone in which asexual reproduction is inhibited (Ilarionova, 2006). Gametogenesis occurs within each gametangium to produce haploid nuclei with *n* number of chromosome. Two haploid nuclei from each gametangia fuse via a process known as karyogamy and a diploid thick-walled oospore with a nucleus is formed. The offspring of one of the mating types then starts developing from the germinated oospore to form a sporangia and the asexual cycle starts all over again (MetaPathogen, no date) The oospores have a dual role, that is, they act as tough structures that resist adverse environmental conditions and they also produce genetic variation by sexual recombination.

The presence of sexual reproduction leads to a more diverse population of *P. infestans* at the genetic level, as a result of genetic mitotic recombination (Cooke & Lees, 2004). Thus, the emergence of A2 mating type coupled with the appearance of 'new' aggressive A1 population has subsequently resulted in an increased genetic diversity which is closely associated with an increased level of resistance to metalaxyl fungicide (Mazakova *et al.*, 2006).

2.4 Symptoms of Late Blight

Phytophthora infestans damages the foliage, stems and tubers of potato plants (*Solanum tuberosum*). The initial symptoms of late blight appear as small, water soaked spots , often with a chlorotic halo, on the stems and leaves (The Pennsylvania State University, 1998). As the disease progresses, necrotic lesions enlarge into a purple-black colour, which later spreads across the whole leaves to the petioles and the stem.

Zones of white masses of sporangia form on the abaxial side of leaves which is visible to the naked eyes (Birch & Whisson, 2001). As the infected leaves die, the aerial parts rot away giving a characteristic odour. In dry conditions, the lesions stop enlarging, darken and wither. The sporangia may detach and fall onto the tubers. The infection of the tuber in its early stages presents slightly brown or purple patches on the skin and there is a rapid decaying of the tuber before harvest (Birch & Whisson, 2001). When the infection of the tuber occurs, there is also the initiation of secondary fungal or bacterial invasion which is known as 'wet rot' (Birch & Whisson, 2001).

The white border of the lesion on the abaxial surface of the leaf contains sporangia of *P.infestans*

Figure 2.4: Symptoms of late blight on the abaxial side of potato leaf (Picture taken and modified by Nesaratnam Alwar; Mare-Longue, 11th September 2014).

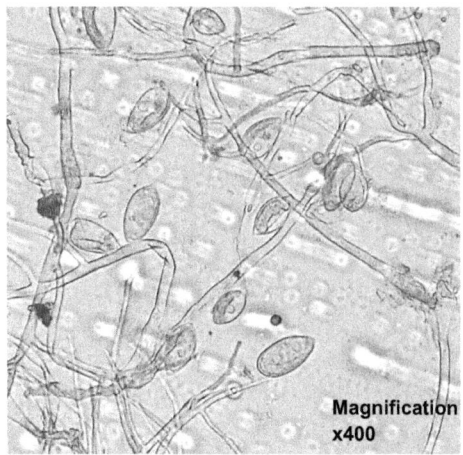

Magnification x400

Figure 2.5: Lemon-shaped sporangia attached to sporangiophores, stained with lactophenol blue solution (isolate from Mare-Longue) (Picture taken by Nesaratnam Alwar with Leica DMIL LED Inverted Trinocular microscope with high definition camera; Biosciences Laboratory, University of Mauritius, 12th September 2014).

10

2.5   Molecular Markers and Genetic Fingerprinting

Molecular markers can measure genetic relatedness more accurately than other types of markers, namely, morphological and biochemical. Molecular markers are simply indicators of sequence polymorphism among individuals which can be due to multiple bases being inserted or deleted, or it can be due to single nucleotide polymorphisms (Edwards & Mogg, 2001, p. 1; Brookes, 1999). Molecular markers do not possess any significant biological effect but they act as landmarks in the genome and are linked to or are part of a gene (Semagn *et al.*, 2006). They are inherited from one generation to the next. Knowledge of the position of markers on a chromosome and their proximity to genes that code for desirable traits is of economic and agronomic importance. There are various molecular characterization tools that differ in their principles and a thorough consideration is therefore required when choosing a molecular marker for genetic studies (Roychowdhury *et al.*, 2013). The choice of marker may ultimately depend on the type of application, the availability of laboratory facilities, expertise, time and a suitable budget.

2.5.1 RAPD Fingerprinting

Progresses in molecular biology techniques have made a considerable number of highly useful DNA markers available for the detection of genetic polymorphism (Bardakci, 2001). During the previous decade, the Random Amplification of Polymorphic DNA (RAPD), which is a variant of the Polymerase Chain Reaction (PCR), has been extensively used to develop DNA markers (Kumar & Gurusubramanian, 2011). The RAPD markers are anonymous DNA sequences which have been amplified at random in a thermocycler and which make use of single, short oligonucleotide primers and therefore, it is not required to have a prior knowledge of the DNA sequence being studied (Bardakci, 2001). The amplified DNA sequence with primers is primarily formed by the interaction between DNA polymerase, RAPD primer and template annealing sites (Semagn *et al.*, 2006). In comparison, Restriction Fragment Length Polymorphism (RFLP) assay which requires restriction enzyme digestion and connected with DNA hybridization is a slow process (Bardakci, 2001). The widespread use of RAPD is due to the benefit of generating a considerable amount of genetic markers from only small amounts of DNA, therefore, any other forms of molecular characterization such as cloning or sequencing of the genome of the species being examined is not required (Gupta *et al.*, 2010). Nonetheless, due to the fact that DNA amplification by RAPD is of a random nature and generated from primers at random, it is fundamental to optimize and preserve constant reaction conditions in order to develop a standard protocol.

2.5.2 Process of RAPD fingerprinting

STEP 1:   DNA is isolated from the sample of cultured *P.infestans*.

STEP 2:   It involves the PCR amplification of the DNA with the primers and dNTPs. The denaturing of the DNA takes place to produce a single strand of DNA. It is followed by the annealing of the RAPD primers and their extensions.

The idea of RAPD technique is that, a short primer that binds to different loci on the genomic DNA, is used for amplification of random sequences. At low annealing temperatures, the primers which are short sequences anneal to their complementary sequences on both DNA strands. Even though arbitrary primers are used, two important rules should be considered: the use of primers with a minimum of 40% GC content and the absence of palindromic sequence (Williams *et al.*, 1990).

STEP 3: When the amplified DNA bands (0.5–5 kb size range) obtained after gel electrophoresis using ethidium bromide staining, are observed with the help of UV light, specific banding patterns known as genomic fingerprints are produced as stated by Jones *et al.* (1997, cited in Kumar *et al.*, 2009). Also, among the amplified DNA bands obtained, are the ones that may be amplified from some genomic DNA of a particular organism only but not from others which therefore indicates that the presence or absence of the amplified fragment is polymorphic in the population of the organism tested.

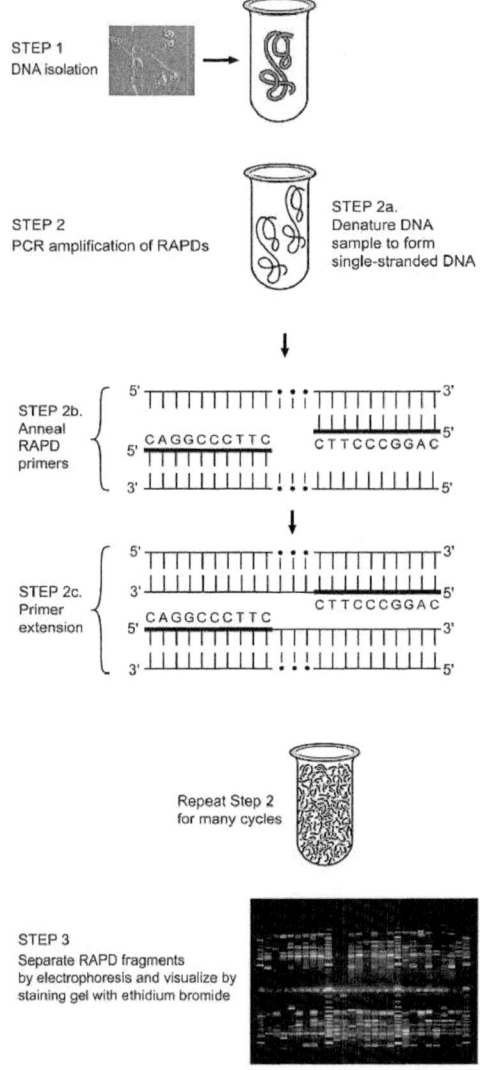

STEP 1
DNA isolation

STEP 2
PCR amplification of RAPDs

STEP 2a.
Denature DNA
sample to form
single-stranded DNA

STEP 2b.
Anneal
RAPD
primers

5' CAGGCCCTTC

CTTCCCGGAC 5'

STEP 2c.
Primer
extension

CTTCCCGGAC

CAGGCCCTTC

Repeat Step 2
for many cycles

STEP 3
Separate RAPD fragments
by electrophoresis and visualize by
staining gel with ethidium bromide

Figure 2.6: Summary of the steps involved in RAPD fingerprinting (White
et al., 2007, p.67) (Modified by Nesaratnam Alwar).

2.5.3 Optimization of RAPD-PCR Method

A few key factors need to be taken into consideration concerning the optimization of the RAPD process in order to achieve a standardised protocol. Quality of the result for the RAPD-PCR method depends on an array of variables. RAPD-PCR is an enzymatic reaction and is laboratory-dependent (Kumar & Gurusubramanian, 2011). The following key variables must be carefully controlled: the quality and concentration of the template DNA; the quality of PCR components used; primer size and the percentage GC content; PCR cycling parameters (more specifically the annealing temperature); the concentration of Taq DNA polymerase enzyme; the concentration of Magnesium chloride; the PCR cycling conditions and the type of thermocycler used (Kumar & Gurusubramanian, 2011). The careful standardization of the technique and reagents is hence required.

2.5.4 DNA template concentration in RAPD fingerprinting

An efficient protocol for RAPD fingerprinting should be resistant to variations in concentration of the DNA template (Skoric et al., 2012). RAPD amplification of DNA fragments does not take place under a certain amount of DNA template concentration .It is also vital to take into consideration that amplification of DNA fragments in this process can be inhibited by high genomic DNA concentrations or due to the presence of PCR inhibitors remaining after DNA extraction (Dias-Neto et al., 1993). In order to improve the RAPD fingerprinting method, it is important to obtain good quality, high molecular weight DNA. The DNA should be quantitated and tested for reproducibility of profiles with at least a two-fold dilution and two-fold concentration of the optimum.

2.5.5 Applications of RAPD-PCR for genetic characterization of *P.infestans*

RAPD fingerprinting has been used worldwide since the last two decades for the genetic characterization of *P.infestans* populations. A study by Atheya et al. (2005) using RAPD in India and Himalayan hill was focused on the level of genetic diversity of populations of *P.infestans* occurring either on hills or plains. Similarly, RAPD analysis of *P.infestans* isolates in China evaluated the relationship among isolates occurring in different provinces (Xiao Qiong et al., 2006). Outbreaks of late blight disease in Turkey during the year 1997 led to a survey carried out between the year 1999 and 2000 where 25 isolates from different locations had been genetically characterized using 21 RAPD primers (Yildirim et al., 2007).

Chapter 3
Methodology

3.1 Overview of Methodology

3.1.1 Protocol used

The protocol used for DNA extraction was adapted from the online 'Laboratory Manual for *P. infestans* work at CIP' by Fry *et al.* (2007) which is accessible from the following website:

https://research.cip.cgiar.org/typo3/web/fileadmin/icmtoolbox/ICM_Toolbox/Files/Manual_draft1.pdf

3.1.2 Outline of the Methodology for this project

The methodology of this work is divided into five main steps:

1) Collection of infected leaves on the site of Mare-Longue.
2) Culture of the different *P.infestans* isolates obtained from infected leaves and from already available cultures on Rye B medium.
3) DNA extraction from the cultured isolates.
4) Screening of 30 primers
5) RAPD fingerprinting using different concentrations of DNA template.

3.1.3 Sources of the isolates

The first isolate was obtained from a potato field at Mare-Longue and was named 'MLo'. Existing cultures of the fungus were obtained from the Biosciences Laboratory and were named 'T11', 'TS1', 'PS1', 'R1' and 'StP'. T11 and 'TS1' were originally isolated from tomato. 'PS1', 'R1' and 'StP' were previously isolated from potato.

3.2 Isolation of P.infestans strains from the field

Infected potato leaves were collected at Mare-Longue on the 11th September 2014. The infected potato plants were from the variety *Delaware* and *Spunta*. The infected leaves were excised and placed in Ziploc bags along with moist pieces of cotton wool. The moist condition was required to preserve the oomycete due to the hot and dry climate on that day. The infected leaves were incubated at 15°C overnight until fresh sporulation appeared.

Isolation:

1. The next day, potatoes of the varieties *Delaware* and *Spunta* were thoroughly washed and cut into slices of about 1cm thick.
2. In a completely sterilised laminar flow hood, the sporulating borders of lesions were cut using a pair of sterile scissors (dipped in 90% alcohol and flamed).
3. The pieces of infected leaves were then placed in Petri dishes containing filter papers which have been dampened with sterile distilled water.
4. A piece of potato disc was then placed on the leaves. It is recommended to put the infected tissues underneath the slice of potato as it favours sporulation.
5. 16 petri dishes containing the *Delaware* potato variety and 16 petri dishes containing the *Spunta* potato variety were labelled accordingly and sealed with PARAFILM M film. The petri dishes were then incubated at 17°C. The incubation lasted for 3 weeks until there were enough sporulation on the upper side of the potato slices.

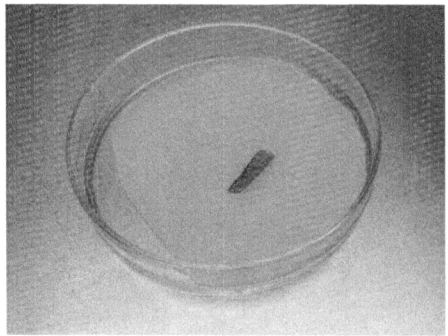

Figure 3.1: Part of the infected leaf containing sporulating lesions. (Picture taken by Nesaratnam Alwar; Biosciences Laboratory, University of Mauritius, 12th September 2014).

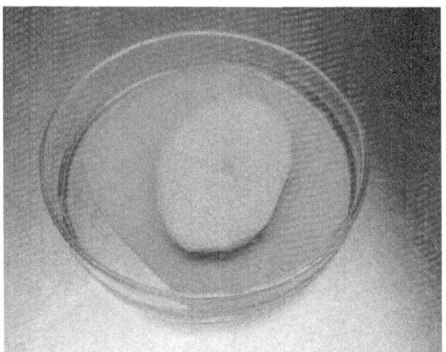

Figure 3.2: Slice of potato placed on infected tissue. (Picture taken by Nesaratnam Alwar; Biosciences Laboratory, University of Mauritius, 12th September 2014).

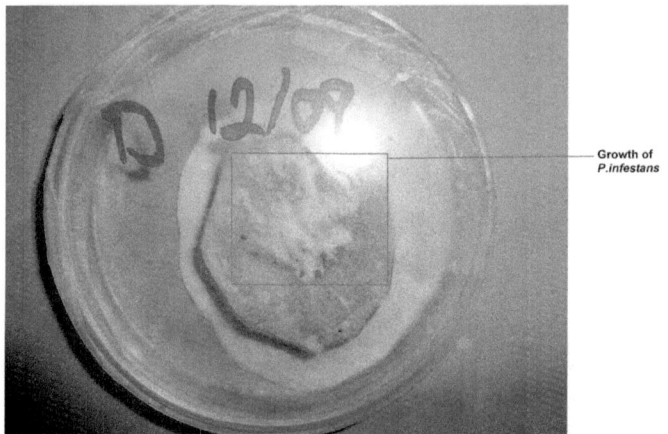

Growth of
P.infestans

Figure 3.3: Growth of *P.infestans* after 3 weeks (Picture taken by Nesaratnam Alwar; Biosciences Laboratory, University of Mauritius, 3rd October 2014).

3.3 Preparation of Cornell medium

The following antibiotic stock mix was prepared with 10ml Dimethyl sulfoxide (DMSO) and aliquoted in 1 ml Eppendorf tubes. The Eppendorf tubes were then stored in the freezer at -20°C. 1ml of the antibiotic mix was to be used per litre of agar being prepared for sub culture. The antibiotic mix has a brick red colour.

Table 3.1: The composition of the antibiotic mix added to 1 litre of Rye B Agar medium.

Antibiotic	Final concentration in media	Quantity of antibiotic powder to weigh out to prepare 10 ml of antibiotic stock mix
Vancomycin	100 mg/L	1 g
Polymixin B	50 mg/L	0.5 g
Ampicillin	200 mg/L	2 g
Rifampicin	20 mg/L	0.2 g
PCNB (pure)	50 mg/L	0.5 g
Benomyl (pure)	50 mg/L	0.5 g

(Source: Laboratory Manual for P. infestans work, International Potato Center, 2007)

3.4 Preparation of Rye B Agar medium

3.4.1 Composition of Rye B Agar medium

Rye B Agar is used for sporulation of *Phytophthora infestans*.

Table 3.2: The composition of Rye Agar B medium. Formula adjusted, standardized to suit performance parameters.

Ingredients	grams/litre
Rye	60.0
Sucrose	20.0
Beta-sitosterol	0.05
Agar	15.0

(Source: HiMedia Laboratories Technical Data, 2011)

Rye is a cereal grain which supplies the elements manganese, phosphorous, magnesium and the amino acid tryptophan to the pathogen. Sucrose is the carbohydrate source and Beta sitosterol helps in sporulation of the oomycete.

3.4.2 Preparation Rye Agar B medium

1. 23.76g of Rye Agar B powder was put into 4 conical flasks of 1000ml capacity each. This made a total of 95.05g of Rye Agar B powder.
2. 250ml of sterile distilled water was added to each conical flask using a measuring cylinder.
3. The 2 previous steps were carried out instead of putting 95.05g of Rye Agar B powder in only one conical flask then adding 1 litre of sterile distilled water. This was to ensure that foaming and spillage of the agar did not take place in the autoclave machine.
4. The medium in the 4 conical flasks was stirred to dissolve the light brown hygroscopic soft lumps.
5. Each conical flask was plugged with cotton wool and covered with aluminium foil.
6. They were put in autoclave at 121°C for 20 minutes.
7. In the meantime, the laminar flow hood where the agar would be put in petri dishes, was cleaned with 70% alcohol and 5% javel. A gas cartridge camping stove was lit to provide a sterile environment.
8. After removal from the autoclave, the flasks were left to cool down to about 60°C in the laminar flow hood.
9. The contents of the 4 conical flasks were poured in a sterile conical flask of 1000ml capacity and 1ml of the antibiotic mix was added. The mixture was then stirred until the brick red colour of the antibiotic mix was evenly distributed throughout the agar medium.
10. 75% of alcohol was sprayed on both hands before pouring the agar onto the plates.
11. The molten agar was then poured in 20 plastic petri dishes and left to set for 45 minutes. This also gave time for the moisture on the petri lids to evaporate.
12. The petri dishes were then sealed with PARAFILM M film.
13. The petri dishes were left overnight in the laminar flow hood.

Note: The neck of the conical flask was flamed from time to time for a few seconds. This was not done to kill microorganisms but to produce an upward flow of air from the flask such that any microorganism in the area will not fall in the flask.

Figure 3.4: Rye Agar B medium in petri dishes (Picture taken by Nesaratnam Alwar; Biosciences Laboratory, University of Mauritius, 2nd October 2014).

3.5 Subculture technique

3.5.1 1st Subculture

While waiting for the isolate 'MLo' to fully grow, the five other isolates were subcultured on the agar medium. The process was carried out in the laminar flow hood.

1. A sterile scalpel (dipped in 90% alcohol and flamed) was used to collect five small puffs of mycelia from the plates and putting them on agar. Care was taken not to scrape off the potato tissues or the agar medium of the cultured isolate.
2. The petri dishes were then sealed with PARAFILM M film and incubated and 17°C.

   The scalpel is constantly dipped in alcohol and flamed especially when culturing various strains so as to avoid cross contamination.

3.5.2 2nd Subculture

After 2 weeks of maintenance of the subcultures and removal of contaminants, a 2nd subculture was made which involved all the six isolates: 'MLo', 'T11', 'TS', 'PS', 'R1' and 'StP'.

3.6 DNA Extraction

1. The Petri with the purest culture was chosen and mycelia were harvested by gently scraping the aerial portion surface of the agar plate.

2. Approximately 100mg of the mycelia was weighed using an electronic balance and it was put in a mortar. 1ml of extraction buffer was added and the mycelia were grounded until fine particulates were obtained. This was done to break the cells and release the nucleus.

3. Using a micropipette, the mixture was collected from the mortar and put in 1ml Eppendorf tubes. They were incubated for 1 hour at 65 °C. Once or twice during that time, the contents were gently mix by inverting the tubes.

4. 333μl of Potassium acetate was added and the tubes shaken vigorously and put on ice for 20 minutes.

5. The tubes were spun at 14,000rpm in the Hettich MIKRO 200R Centrifuge for 10 minutes. This was the last step in the preparation of cell extract and it involved the formation of a pellet consisting of cell debris and partially digested organelles and leaving the cell extract as a clear supernatant.

6. The Eppendorf tubes then contained an aqueous clear layer containing DNA and grayish pellet which consisted of cell debris. The aqueous supernatant was removed carefully using a micropipette and put into a sterile 2ml Eppendorf tubes. Care was taken not to include the grayish pellet.

7. Purification of DNA from cell extract: 800μl of cold isopropanol was added, the tubes were inverted to mix the content and the tubes were put on ice for 30 minutes. This step favours nucleic acid precipitation.

8. The tubes were then centrifuged at 14,000rpm for 5 minutes. The DNA was precipitated as pellets and the supernatant was discarded. The pellets were dried by placing the tubes on a heat block.

9. For a 2nd precipitation, the pellets were resuspended in 700µl of TE buffer.

10. 75µl of 3M Sodium acetate and 500µl of isopropanol were then added. The contents were mixed by inversion and spun down for 30 seconds. The tubes are then stored at -20°C for overnight precipitation.

11. After overnight precipitation, the tubes were centrifuged at 13,000rpm for 30 minutes.

12. The supernatants were dumped and the pellets were washed with 75% ethanol and centrifuged again at 13,000rpm for 10 minutes.

13. The supernatants were discarded and the pellets were dried by placing the tubes on a heat block.

14. The pellets were resuspended in 50µl of TE buffer and stored at -20°C.

3.6.1 Solutions used in DNA extraction of *Phytophthora infestans*

It is important to note that when preparing the Extraction and TE buffer, all of their components were 10 times more concentrated and kept as stock solutions.

1. **Extraction buffer**

Table 3.3 The composition of Extraction buffer.

Buffer	Chemical	Final Concentration
	EDTA	0.05M
	Tris pH 8.0	0.1M
Extraction	NaCl	0.5M
	Beta mercaptoethanol	0.7%
	SDS	0.25%

(Source: Laboratory Manual for P. infestans work, International Potato Center, 2007)

The Extraction buffer helps in maintaining the structure of DNA during breakage and purification steps. It also causes the inactivation of DNA degrading enzymes present in the cells.

23

Components of the Extraction buffer

- **Ethylenediaminetetraacetic acid (EDTA)**

  EDTA is a chelating agent that binds with Mg^{2+} which is an essential cofactor of DNases, thereby inhibiting the activities of the enzymes (Allers & Lichten, 2000).

- **Sodium dodecyl sulfate (SDS)**

  SDS breaks down the cell membrane by emulsifying lipids and denaturing proteins. This hinders the polar interactions occurring in the membrane (Simon Fraser University (sfu), no date).

- **Sodium chloride (NaCl)**

  NaCl increases the stability of the solution as Na^+ forms a cloud of positive charges around the DNA more specifically, it shields the negative phosphate ends of the DNA strand. This causes the strands to regroup and the nucleic acid to precipitate out of organic solutions.

- **Beta-Mercaptoethanol (ß-ME)**

  Beta-mercaptoethanol (ß-ME) is a reducing agent that breaks disulfide bonds and causes the denaturation of RNases, hence inhibiting the activities of the enzyme (QIAGEN - Sample & Assay Technologies, no date).

- **Tris pH 8.0**

  Tris protects the DNA from changes in pH

2. **TE buffer**

Table 3.4: The composition of TE buffer.

Buffer	Chemical	Final Concentration
TE	Tris HCl pH 8.0	10mM
	EDTA	1mM

(Source: Laboratory Manual for P. infestans work, International Potato Center, 2007)

3. **Isopropanol**

Isopropanol is used to precipitate DNA.

4. **Ethanol**

Ethanol is particularly a better choice when carrying out DNA precipitation. Often, ethanol is used if small volumes of DNA are being precipitated. In this way, larger volumes of DNA can be recovered without having to worry about salt contamination as when using isopropanol because the salt remains soluble even at low temperatures in ethanol (Molecularcloning, 2012).

3.6.2 DNA Quality & Concentration investigated by Gel Electrophoresis

1. **Preparation of TBE buffer (x10)**

Reagents needed:

10.8g Tris base

5.5g Boric acid

4ml 0.5 EDTA (pH 8)

100ml sterile distilled water

10.8g Tris base, 5.5g Boric acid and 4ml 0.5 EDTA (pH 8) were added to 100ml sterile distilled water to prepare a stock solution of TBE buffer (x10).

Dilution procedure:

The TBE buffer was then diluted to x1 by adding 20ml of the stock solution (x10) to 180ml of sterile distilled water.

2. **Preparation of small size 1.5% agarose gel**

0.75g agar powder was dissolved in in 50ml of 1 x TBE buffer in a conical flask. The mixture was heated for about 45-50 seconds in a microwave. It was swirled and heated again until a clear solution was obtained. A few drops of Ethidium bromide was then pipetted in the conical flask.

The agar was left to cool on the bench and then it was poured in a gel tray. Immediately, the comb was inserted and the gel was left to set for 15 minutes.

3. Preparation of DNA samples of the isolates

5µl of DNA sample from an isolate was put in 0.2ml Eppendorf tube. 2µl of 6 x DNA Loading Dye (Fermentas) was added to the DNA sample. The process was repeated for other DNA samples.

4. Loading of the agarose gel

Procedure:

→ The agarose gel was put in the tank and 1 x TBE buffer was poured over the gel up to the graduated level in the tank.

→ 3µl of 1kb DNA Ladder (GeneRuler™, Fermentas) was loaded in the well first and the remaining wells were loaded with 7µl of DNA samples each. The 1kb DNA Ladder (GeneRuler™, Fermentas) is ideal for both DNA sizing and approximate quantification. Care was taken not to damage the well with the tip of the micropipette.

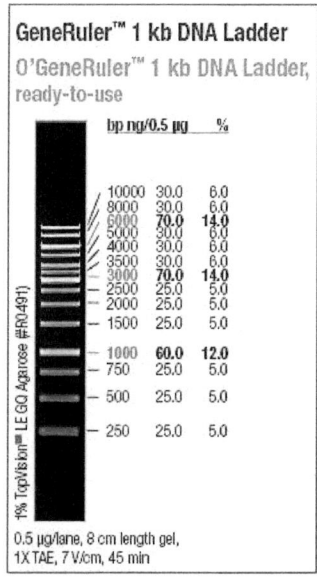

Figure 3.5: GeneRuler 1kb DNA ladder, ready-to-use (Source: Thermo Scientific, no date).

→ A few drops of Ethidium bromide was added to the running buffer.

→ The tank was closed and the gel was run at 100V for 1hour and 30 minutes.

→ After the running time, the gel was visualized under UV transillumination.

℘ **Precautions** should be taken when dealing with Ethidium bromide as it is a suspected carcinogen. Additionally, after the electrophoresis, the gel and the buffer were discarded in special waste bucket.

3.6.3 Determination of DNA concentration by spectrophotometric estimation

An easy way of determining the concentration of DNA is by spectrophotometric analysis. The bases present in the DNA strand absorb UV light, therefore the concentration of the DNA solution is positively correlated to the amount of UV light being absorbed.

Rule: Concentration of pure double-stranded DNA with an A_{260} of 1.0 = 50 µg/ml

The following formula can be used to determine the DNA concentration of a solution (Promega, no date).

Unknown ng/µl (equivalent to µg/ml) = 50 ng/µlx Measured A_{260} x dilution factor

1. The spectrophotometer was blanked with 1 ml sterile TE buffer first.
2. 5 µl of the DNA preparation from an isolate was diluted with 995 µl sterile TE buffer (dilution factor x 200) and readings were taken at 260 nm and 280 nm respectively.
3. Triplicate readings for each isolate were taken at each wavelength and a mean value of DNA concentration was worked out.
4. The purity of the DNA was evaluated using the A_{260}/A_{280} ratio.
5. The concentration of DNA was calculated using the above formula (in bold character).

3.7 Screening of Primers

Before testing different concentrations of DNA, the primers (Operon Technologies Inc., CA, USA) needed to be screened and those giving the best results were selected. For this project work, 30 primers were screened. RAPD primers act as both forward and reverse decamers.

Table 3.5: List of primers that were screened (Operon Technologies Inc., CA, USA)

Primer name	Primer sequence	Number of bp	% G-C content
OPB-01	GTTTCGCTCC	10	60
OPB-02	TGATCCCTGG	10	60
OPB-03	CATCCCCCTG	10	70
OPB-04	GGACTGGAGT	10	60
OPB-05	TGCGCCCTTC	10	70
OPB-06	TGCTCTGCCC	10	70
OPB-07	GGTGACGCAG	10	70
OPB-08	GTCCACACGG	10	70
OPB-09	TGGGGGACTC	10	70
OPB-10	CTGCTGGGAC	10	70
OPE-01	CCCAAGGTCC	10	70
OPE-02	GGTGCGGGAA	10	70
OPE-03	CCAGATGCAC	10	60
OPE-04	GTGACATGCC	10	60
OPE-05	TCAGGGAGGT	10	60
OPE-06	AAGACCCCTC	10	60
OPE-07	AGATGCAGCC	10	60
OPE-08	TCACCACGGT	10	60
OPE-09	CTTCACCCGA	10	60
OPE-10	CACCAGGTGA	10	60
OPL-01	GGCATGACCT	10	60
OPL-02	TGGGCGTCAA	10	60
OPL-03	CCAGCAGCTT	10	60
OPL-04	GACTGCACAC	10	60
OPL-05	ACGCAGGCAC	10	70
OPL-06	GAGGGAAGAG	10	60
OPL-07	AGGCGGGAAC	10	70
OPL-08	AGCAGGTGGA	10	60
OPL-09	TGCGAGAGTC	10	60
OPL-10	TGGGAGATGG	10	60

3.8 Storage and manipulation of the DNA templates

The DNA of the isolates, each of a capacity of 30µl (remaining after spectrophotometric analysis and gel electrophoresis) were stored at -20°C. Storing DNA at high concentration prevents its degradation. When required, a certain amount of the DNA from each isolate was taken and diluted in TE buffer. Each time, a range of concentration was tested and the clarity of the bands were evaluated.

3.9 RAPD fingerprinting

The following stock solutions were first set up:
- → 50 µl of 2.5mM dNTPs (Fermentas)
  (a mix containing 2.5 mM dATP, 2.5 mM dCTP, 2.5 mM dGTP, 2.5 mM dTTP)
- → 50 µl of 10 µM primer (Operon Technologies Inc., CA, USA)
- → Sterile distilled water

Each RAPD-PCR reaction should **ALWAYS** have the following 6 components:

1. Sterile distilled water
2. 5 x PCR buffer (containing $MgCl_2$) (Thermo Scientific)
3. dNTPs (Fermentas)
4. RAPD primer (acting as both forward and reverse primer)
5. Taq polymerase (Thermo Scientific)
6. DNA template

Here we varied the concentration of the DNA template only

The master mix was prepared in stock with one extra unit amount (to compensate for pipetting errors) and then it was distributed to the respective 0.2ml Eppendorf tubes. The appropriate RAPD primers and DNA template were added to the respective tubes and the contents were mixed.

Table 3.6: Composition of PCR Master mix

Component	Amount in µl per tube	Final concentration in one tube
Sterile distilled water	7.8	-
5X PCR buffer	5.0	1X
dNTPs (2.5 mM)	2.0	0.2 mM
RAPD primer (10 µM)	5.0	2.0 µM
Taq polymerase	0.2	0.2 units
DNA template (ng/µl)	5.0	?
TOTAL VOLUME	25	

Table 3.7: PCR Cycling conditions

Step	Temperature (°C)	Time	Number of cycles
Initial denaturation	94	2 mins	1
Denaturation	94	30 sec	
Annealing	35	30 sec	30
Extension	72	1 min	
Final extension	72	5 mins	1
Hold	10	12 hrs	

The PCR products were then run on a 2% agarose gel at 80V for 1 hour and 30 minutes. A 50bp DNA Ladder was used.

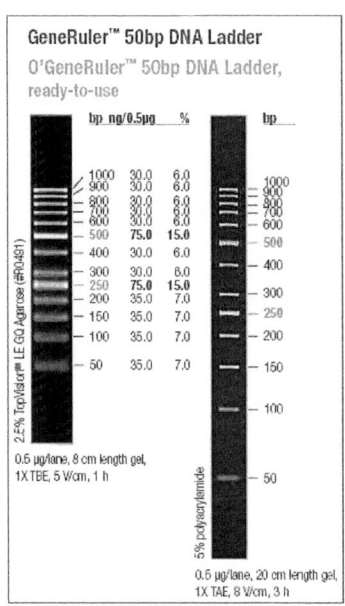

Figure 3.6: GeneRuler 50 bp DNA Ladder, ready-to-use (Source: Thermo Scientific, no date).

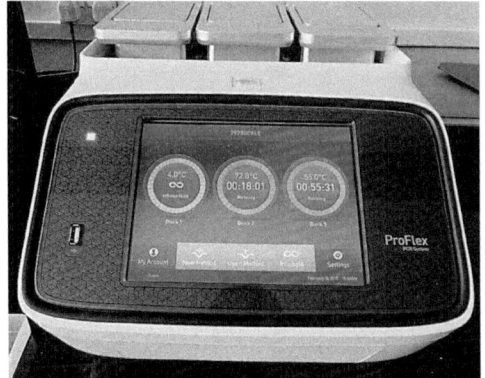

Figure 3.7: The ProFlex™ PCR System is a thermal cycler that has a triple block (3 x 32 wells). (Picture taken by Nesaratnam Alwar; Mare-Longue, 6th February 2015).

Chapter 4
Results

4.1 Culture of *Phytophthora infestans* isolates

The isolates 'T11' and 'PS1' grew very well with minimal to no contamination at all. The isolates 'StP' did not grow at all or were heavily contaminated. The isolates 'R1' and 'TS1' grew abundantly on some plates but were spoiled on the majority of the other plates. The isolate 'MLo' which was collected on field at Mare-Longue grew much better on potato tissue than on Rye B Agar medium.

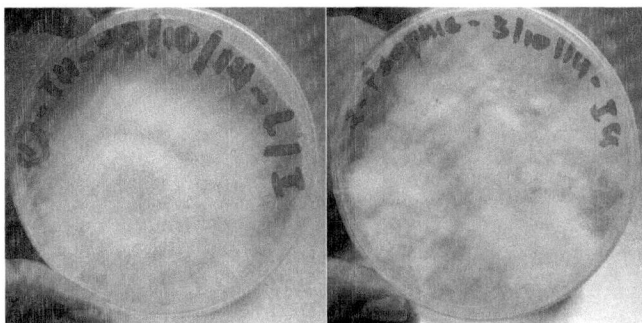

Figure 4.1: Isolates 'T11' and 'PS1' after 3 weeks of culture on Rye B Agar medium.

4.2 Estimation of presence of DNA by fluorescence of Ethidium bromide

Agarose gel electrophoresis is a way to quickly estimate the presence of DNA after DNA extraction. The intensity of fluorescence of Ethidium bromide (ethidium bromide intercalates itself within the DNA) under UV light after electrophoresis was used to estimate the amount of DNA obtained and the DNA quantity was estimated by comparing its level of fluorescence with the intensity of fluorescence of the DNA ladder. This method may not be accurate but it gives a good indication about the quality of the product obtained.

The products of DNA extraction fluoresced much more than the DNA ladder. This gave an indication that a high proportion of DNA was present. (But since the exact values of concentration were calculated from the spectrophotometric values, we did not rely on the fluorescence method much. It only served as an indicator.)

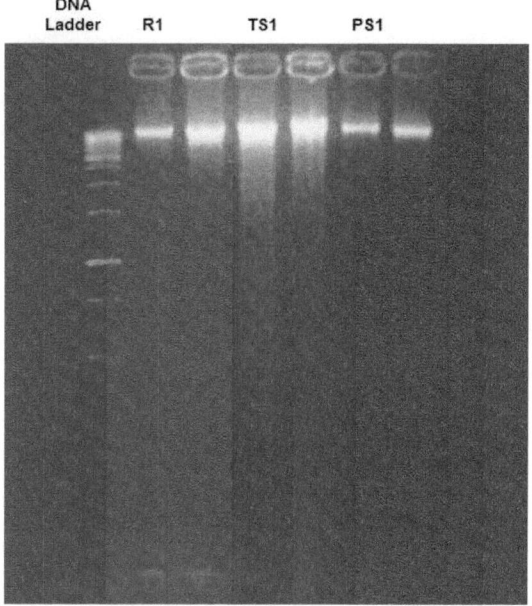

Figure 4.2: Picture of gel showing an estimation of the amount of DNA yielded from DNA extraction.

4.2.1 Result of Spectrophotometric analysis

DNA extraction was carried out on isolate PS1, R1 and TS1. Triplicate readings for each isolate were taken at wavelength A_{260} and A_{280} and the mean value (obtained from the two closest values) was used to calculate the purity and concentration of the DNA yielded.

Table 4.1: Triplicate readings and mean value from spectrophotometric analysis of the DNA.

Isolate	A_{260}			Mean	A_{280}			Mean
TS1	**0.253**	0.227	**0.247**	0.2500	**0.140**	0.121	**0.135**	0.1375
R1	**0.555**	**0.564**	0.539	0.5595	**0.290**	0.301	**0.283**	0.2865
PS1	**0.100**	**0.105**	0.073	0.1025	**0.053**	0.063	**0.057**	0.0550

(Note: In order to calculate the mean values, only the 2 closest values (in bold) from the triplicate readings were taken.)

4.2.2 Purity of DNA

The purity of the DNA was assessed by the ratio A_{260}/A_{280}. A value between 1.8 and 1.9 is generally regarded as "pure" for double stranded DNA as stated by Sambrook *et al.* (1989, cited in Ejaz *et al.*, 2014). The DNA obtained from the isolates was hence considered as pure since the ratios were neither to low nor high.

Table 4.2: Purity of the DNA from the isolates.

Isolate	A_{260}/A_{280}
TS1	1.81
R1	1.95
PS1	1.86

4.2.3 Concentration of DNA yielded

The concentration of DNA for each isolate was calculated using the formula:
Unknown (ng/µl) = 50 ng/µl x Measured A_{260} x dilution factor (Promega, no date) where the dilution factor is 200 since 5µl of DNA was added to 995µl of TE buffer.

Table 4.3: DNA concentration for 3 isolates.

Isolate	50 ng/µl x Measured A_{260} x dilution factor	Concentration (ng/µl)
R1	50 x 0.5595 x 200	5595
PS1	50 x 0.1025 x 200	1025
TS1	50 x 0.2500 x 200	2500

These concentrations of DNA from each isolate act as stock solution. In order to prepare several concentrations of DNA, the stock is then diluted with TE buffer.

4.3 Screening of Primers

All the 30 primers were screened with the isolate R1. All the gel pictures obtained during the laboratory work were analysed and the Rf values of the bands were calculated by using the computer program Thermo Scientific myImageAnalysis Software v2.0 and DNA standard curves were made using Microsoft® Excel 2010.

The screening of primers was also a step where a certain DNA template concentration range was tested. In this way, the best primers but also the range of DNA concentrations to be analysed were obtained.

4.3.1 Measurement of Rf values

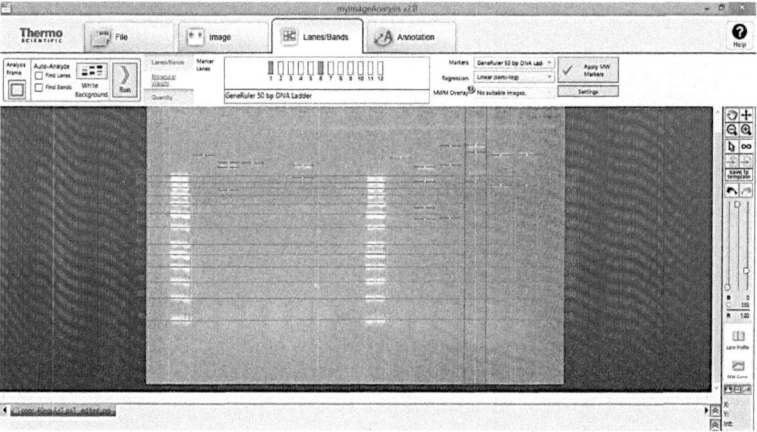

Figure 4.3: Thermo Scientific myImageAnalysis Software v2.0 is a program that measures the Rf values of the DNA bands.

Thermo Scientific myImageAnalysis Software v2.0 has a user-friendly graphical interface and enables a rapid detection of the bands and the calculation of the Rf values for each of the bands present. It also has preset calibrations for every DNA Ladders such that there is no need for the user to insert each amount of base pairs manually. This software also allows the exportation of the data obtained to Microsoft® Excel 2010 to allow a graphical representation of the information. This software was used in this work to increase the accuracy of data.

4.3.2 Generation of Molecular Weight vs Rf value Semi Log Graph

After obtaining the Rf values for the DNA ladder in each picture, these data were entered in Microsoft® Excel 2010 to generate a Molecular Weight vs Rf value semi log graph. The graph is in the Semi-Log form with an exponential trendline passing through the data points. It should be noted that when using the exponential trendline, the R^2 value should be close to 1 (values close to 1 indicate that all points lie approximately on a straight line with minimal scatter. Therefore, knowing the value of x enables the prediction of the value of y). The molecular weights of the bands are then obtained from the equation of the trendline.

4.3.3 Screening of OPB primers with Isolate R1

For the first screening of primers, DNA concentrations of 50,100 and 200 ng/µl were used.

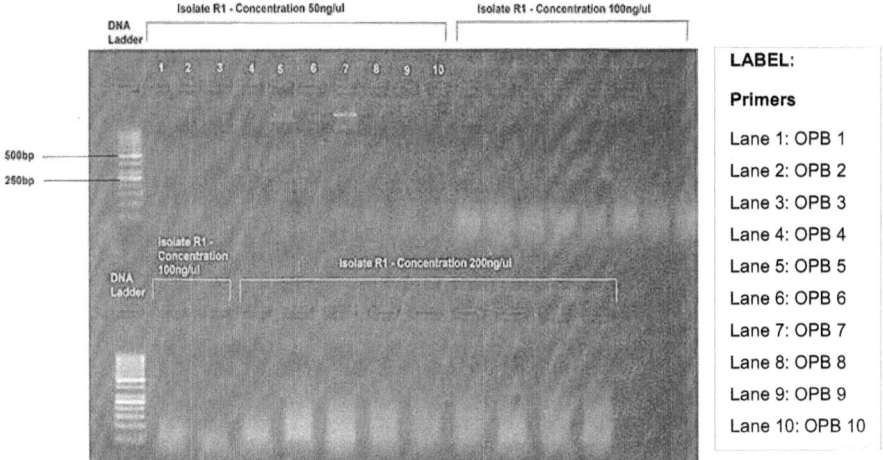

Figure 4.4: Screening of OPB primers at DNA concentrations 50,100 and 200ng/µl. The numbers 1-10 represent the different OPB primers (OPB 1- OPB 10).

Table 4.4: Rf values for the corresponding bands of the 50bp DNA Ladder.

Band (DNA Ladder)	Rf value	Base Pairs (bp)
1	0.31	1000
2	0.338	900
3	0.362	800
4	0.383	700
5	0.408	600
6	0.449	500
7	0.484	400
8	0.544	300
9	0.582	250
10	0.631	200
11	0.676	150
12	0.739	100
13	0.819	50

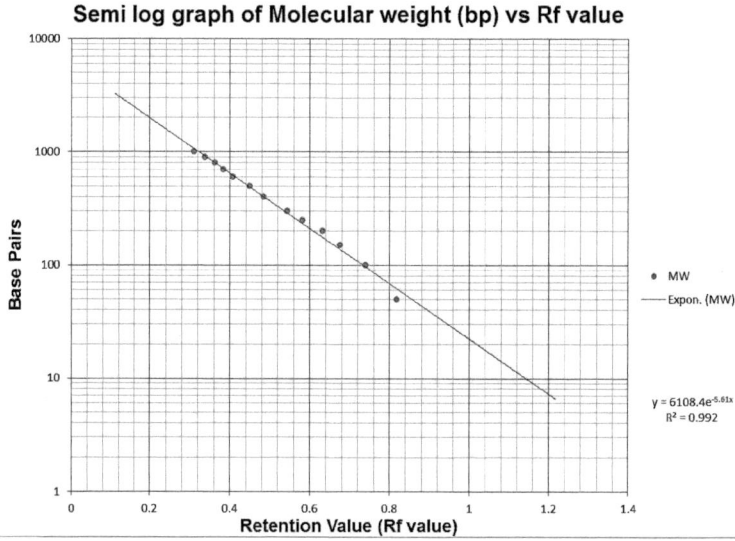

Figure 4.5: Semi log graph of Molecular weight (bp) vs Rf value for the estimation of the molecular weight of the bands obtained with the OPB primers.

From the screening of the OPB set of primers (OPB1 - OPB10) using the R1 DNA template at concentrations 50,100 and 200ng/µl, 2 primers, OPB 5 and OPB 7 gave positive results at concentration 50ng/µl as shown in figure 4.4, table 4.5 and 4.6. The presence of bands was not detected at DNA template concentrations of 100 and 200ng/µl, Instead, smears were produced at those concentrations.

Table 4.5: The number of bands obtained with the OPB 5 primer, and their presence or absence at DNA concentrations 50,100 and 200ng/µl .

OPB 5 Band number	Rf value	Base Pairs (bp)	Presence of bands at the following DNA concentrations		
			DNA concentration 50ng/µl	DNA concentration 100ng/µl	DNA concentration 200ng/µl
1	0.226	1719	✔	x	x

Table 4.6: The number of bands obtained with the OPB 7 primer, and their presence or absence at DNA concentrations 50,100 and 200ng/µl.

OPB 7 Band number	Rf value	Base Pairs (bp)	Presence of bands at the following DNA concentrations		
			DNA concentration 50ng/µl	DNA concentration 100ng/µl	DNA concentration 200ng/µl
1	0.195	2045	✔	x	x
2	0.230	1681	✔	x	x
3	0.268	1358	✔	x	x

4.3.4 Screening of OPE primers with Isolate R1

For the second screening of primers, a DNA concentration of 50ng/µl was used.

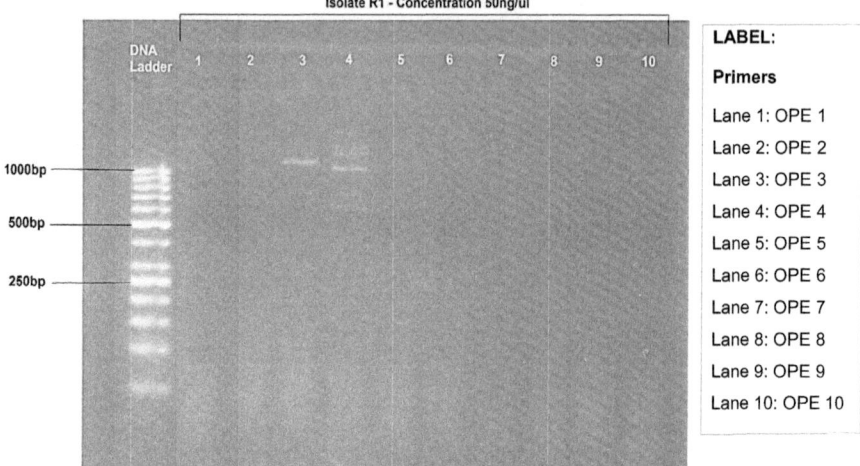

Figure 4.6: Screening of OPE primers at a DNA concentration of 50ng/µl. The numbers 1-10 represent the different OPE primers (OPE 1- OPE 10).

Table 4.7: Rf values for the corresponding bands of the 50bp DNA Ladder.

Band (DNA Ladder)	Rf value	Base Pairs (bp)
1	0.201	1000
2	0.217	900
3	0.238	800
4	0.261	700
5	0.286	600
6	0.320	500
7	0.361	400
8	0.416	300
9	0.452	250
10	0.496	200
11	0.542	150
12	0.603	100
13	0.701	50

39

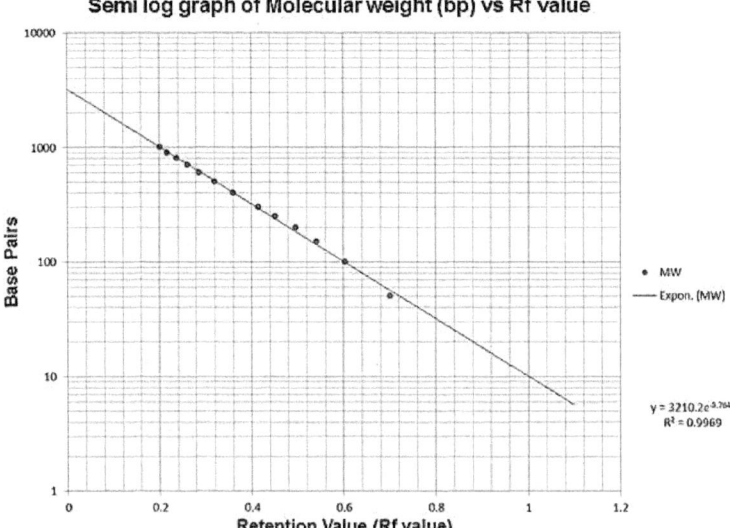

Figure 4.7: Semi log graph of Molecular weight (bp) vs Rf value for the estimation of the molecular weight of the bands obtained with the OPE primers.

From the screening of the OPE set of primers (1-10) using the R1 DNA template at a concentration of 50ng/μl, 2 primers, OPE 3 and OPE 4 gave positive results as shown in figure 4.6, table 4.8 and 4.9.

Table 4.8: The number of bands obtained with the OPE 3 primer, at a DNA concentration of 50ng/μl.

OPE 3 Band number	Rf value	Base Pairs (bp)	Presence of bands at the following DNA concentration
			DNA concentration 50ng/μl
1	0.179	1144	✔

Table 4. 9: The number of bands obtained with the OPE 4 primer, at a DNA concentration of 50ng/μl.

OPE 4 Band number	Rf value	Base Pairs (bp)	Presence of bands at the following DNA concentration
			DNA concentration 50ng/μl
1	0.146	1384	✔
2	0.164	1247	✔
3	0.192	1061	✔
4	0.254	743	✔
5	0.282	632	✔

4.3.5 Screening of OPL primers with Isolate R1

For the third screening of primers, DNA concentrations of 20, 50 and 80ng/µl were used.

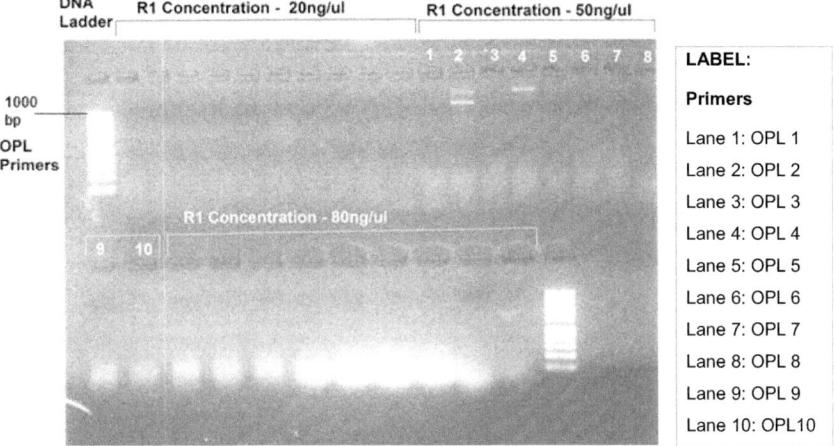

Figure 4.8: Screening of OPL primers at DNA concentrations of 20, 50 and 80ng/µl. The numbers 1-10 represent the different OPL primers (OPL 1- OPL10).

Table 4.10: Rf values for the corresponding bands of the 50bp DNA Ladder.

Band (DNA Ladder)	Rf value	Base Pairs
1	0.298	1000
2	0.319	900
3	0.347	800
4	0.368	700
5	0.397	600
6	0.426	500
7	0.467	400
8	0.527	300
9	0.556	250
10	0.598	200
11	0.645	150
12	0.708	100
13	0.789	50

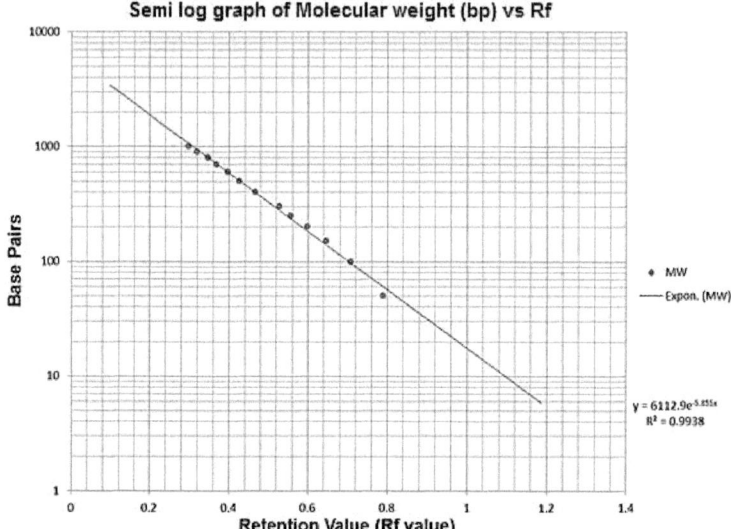

Figure 4.9: Semi log graph of Molecular weight (bp) vs Rf value for the estimation of the molecular weight of the bands obtained with the OPL primers.

The screening of the OPL set of primers (1-10) using the isolate R1 as DNA template at concentrations 20, 50 and 80ng/μl gave rise to two information. Firstly, the OPL 2 and OPL 4 gave positive results during screening of primers as shown in figure 4.8, table 4.11 and 4.12. Secondly, the concentrations of DNA at 20 and 80 ng/μl did not give rise to any band on the agarose gel.

Table 4.11: The number of bands obtained with the OPL 2 primer, and their presence or absence at DNA concentrations 20, 50 and 80ng/μl.

OPL 2 Band number	Rf value	Base Pairs (bp)	Presence of bands at the following DNA concentrations		
			DNA concentration 20ng/μl	DNA concentration 50ng/μl	DNA concentration 80ng/μl
1	0.157	2438	x	✔	x
2	0.225	1637	x	✔	x

Table 4.12: The number of bands obtained with the OPL 4 primer, and their presence or absence at DNA concentrations 20, 50 and 80ng/µl.

OPL 4 Band number	Rf value	Base Pairs (bp)	Presence of bands at the following DNA concentrations		
			DNA concentration 20ng/µl	DNA concentration 50ng/µl	DNA concentration 80ng/µl
1	0.147	2585	x	✔	x

4.4 Testing of DNA Template Concentrations (Part 1)

Once the screening step was completed, an experimental set-up was designed to test which DNA template concentration yielded best results in RAPD-PCR. A concise procedure was followed. Bearing in mind that the RAPD procedure has a problem of reproducibility with the results varying from laboratory to laboratory and day to day, 3 different DNA template concentrations (30, 50 and 70ng/ µl) were tested on the same gel with one primer at a time. Scoring of the bands for each testing was made and the clarity of bands (i.e. band intensity relative to the background) in each lane was assessed on a scale of 1 to 3 with 1 being the best visual appearance and 3 the least.

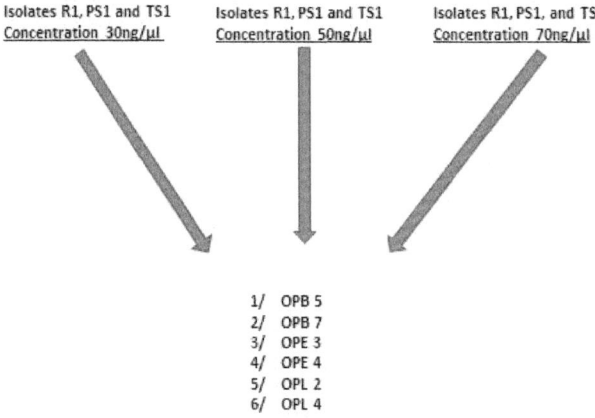

Figure 4.10: Pictorial explanation of how the work was set up. 3 different DNA concentrations of the 3 isolates were tested with a single primer at a time.

4.4.1 Isolates R1, PS1, TS1 at concentrations 30, 50 and 70ng/µl with OPB 5 primer

For the first testing of different concentrations, the OPB 5 primer was used.

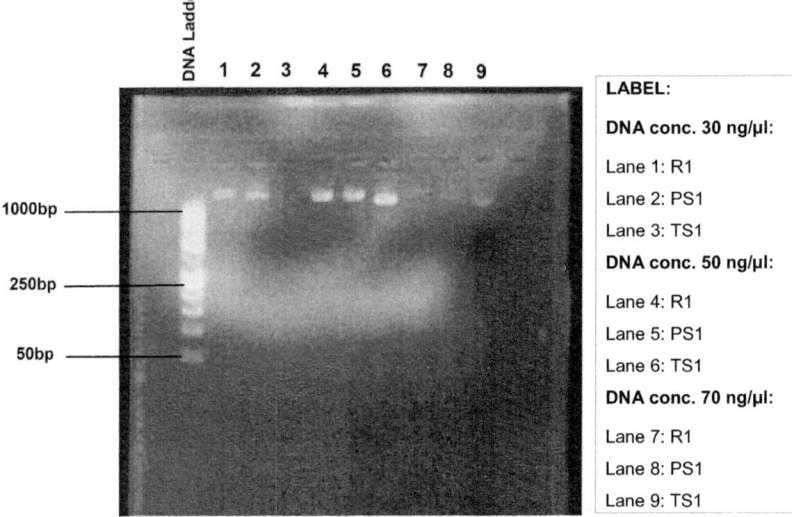

Figure 4.11: RAPD fingerprints of isolates R1, PS1 and TS1 at concentrations 30, 50 and 70ng/µl with OPB 5 primer.

Table 4.13: Rf values for the corresponding bands of the 50bp DNA Ladder.

Band (DNA Ladder)	Rf value	Base Pairs
1	0.240	1000
2	0.253	900
3	0.263	800
4	0.273	700
5	0.291	600
6	0.308	500
7	0.327	400
8	0.352	300
9	0.370	250
10	0.390	200
11	0.414	150
12	0.441	100
13	0.486	50

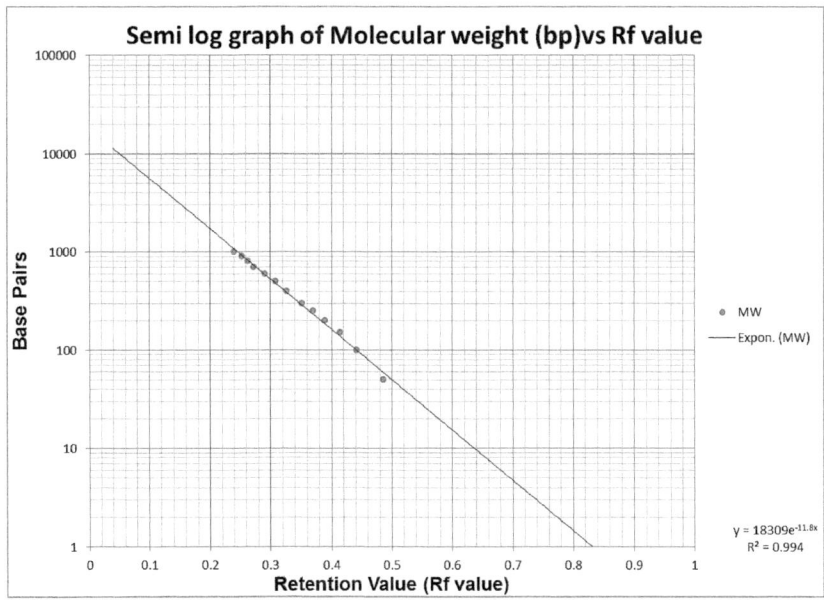

Figure 4.12: Semi log graph of Molecular weight (bp) vs Rf value for the estimation of the molecular weight of the DNA bands obtained at DNA concentrations 30, 50 and 70ng/µl with OPB 5 primer.

Table 4.14: Band Scoring Table for isolates at concentrations 30, 50 and 70 ng/µl with OPB 5 primer.

Rf value	Band (bp)	Concentration 30 ng/µl			Concentration 50ng/µl			Concentration 70ng/µl		
		1. R1	2. PS1	3. TS1	4. R1	5. PS1	6. TS1	7. R1	8. PS1	9. TS1
0.2123	1495	✔			✔			✔		
0.2140	1465		✔			✔			✔	
0.2235	1310						✔			✔
	Total number of bands	1	1		1	1	1	1	1	1
	Clarity	2	2		1	1	1	3	3	3

From figure 4.11 and from table 4.14 it can be seen that a DNA concentration of 50ng/µl gave a good visual appearance with clarity of 1 (clarity being on a scale of 1-3, where 1 is the best appearance and 3 the least, all relative to the background.). At DNA concentration of 30ng/µl, the isolate TS1 did not produce any band. At DNA concentration of 70ng/µl, bands for each isolates were present, however they were slightly faint. Amplified DNA fragments produced bands of size 1310, 1465 and 1495 bp.

4.4.2 Isolates R1, PS1, TS1 at concentrations 30, 50 and 70ng/µl with OPB 7 primer

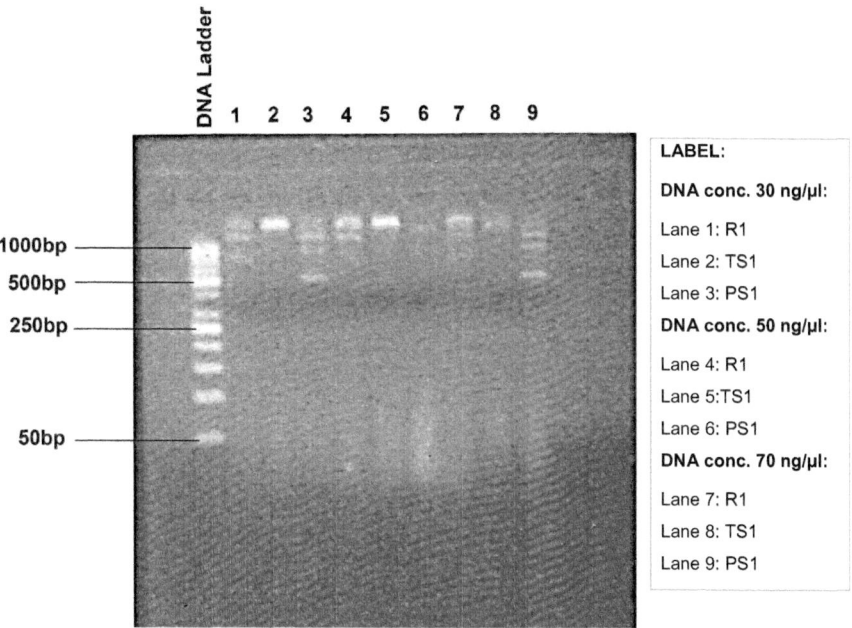

Figure 4.13: RAPD fingerprints of isolates R1, PS1 and TS1 at concentrations 30, 50 and 70ng/µl with OPB 7 primer.

Table 4.15: Rf values for the corresponding bands of the 50bp DNA Ladder.

Band (DNA Ladder)	Rf value	Base Pairs
1	0.197	1000
2	0.203	900
3	0.212	800
4	0.221	700
5	0.230	600
6	0.249	500
7	0.269	400
8	0.299	300
9	0.320	250
10	0.351	200
11	0.384	150
12	0.430	100
13	0.496	50

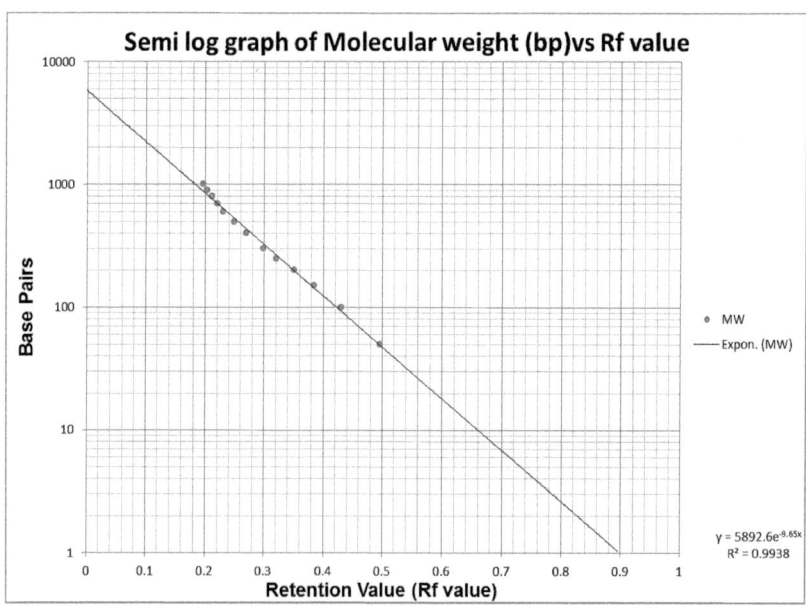

Figure 4.14: Semi log graph of Molecular weight (bp) vs Rf value for the estimation of the molecular weight of the DNA bands obtained at DNA concentrations 30, 50 and 70ng/μl with OPB 7 primer.

Table 4.16: Band Scoring Table for isolates at concentrations 30, 50 and 70 ng/µl with OPB 7 primer.

Rf value	Band (bp)	Concentration 30 ng/µl			Concentration 50ng/µl			Concentration 70ng/µl		
		1. R1	2. TS1	3. PS1	4. R1	5. TS1	6. PS1	7. R1	8. TS1	9. PS1
0.1566	1300	✔	✔	✔	✔	✔		✔	✔	✔
0.1623	1230						✔			
0.1721	1120							✔		✔
0.1772	1065	✔		✔	✔					
0.1947	900			✔			✔			✔
0.2082	790	✔								
0.2325	625									✔
0.2376	595			✔						
	Total number of bands	3	1	4	2	1	2	2	1	4
	Clarity	2	1	2	1	1	3	2	3	1

Firstly, from figure 4.13 and table 4.16, it is seen that bands were present for every isolates and at all concentrations being tested. The size of amplified DNA fragments varied from 595bp to 1300bp. A higher number of bands were visualised with isolate PS1. The isolate TS1 produced only 1 band of 1300bp at all DNA concentrations tested. Based on the number of bands produced and the clarity, DNA concentration of 30 and 50ng/ µl gave better results. DNA concentrations of 30 and 50ng/µl had on overall better clarity.

4.4.3 Isolates R1, PS1, TS1 at concentrations 30, 50 and 70ng/µl with OPE 3 primer

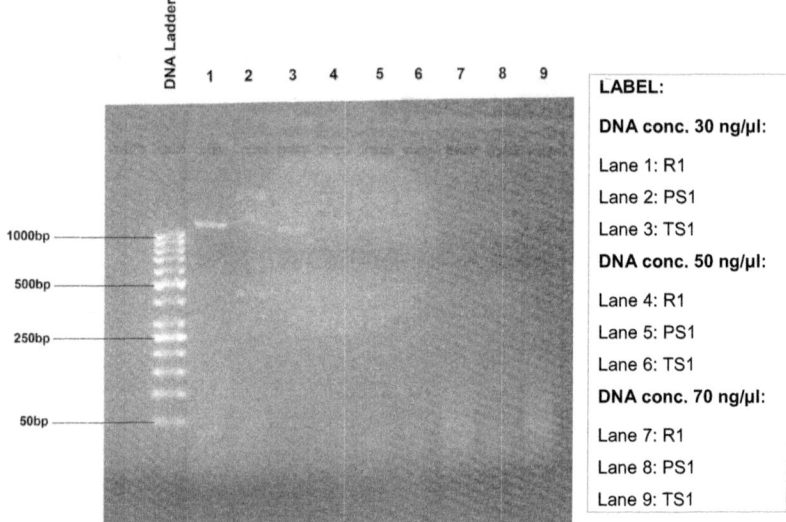

Figure 4.15: RAPD fingerprints of isolates R1, PS1 and TS1 at concentrations 30, 50 and 70ng/µl with OPE 3 primer.

Table 4.17: Rf values for the corresponding bands of the 50bp DNA Ladder.

Band (DNA Ladder)	Rf value	Base Pairs
1	0.277	1000
2	0.287	900
3	0.304	800
4	0.322	700
5	0.340	600
6	0.366	500
7	0.396	400
8	0.436	300
9	0.461	250
10	0.489	200
11	0.523	150
12	0.565	100
13	0.615	50

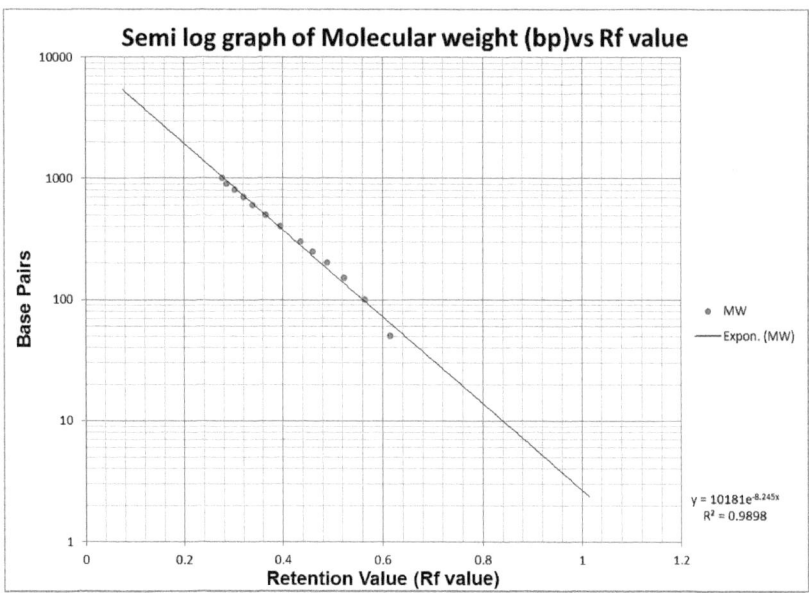

Figure 4.16: Semi log graph of Molecular weight (bp) vs Rf value for the estimation of the molecular weight of the DNA bands obtained at DNA concentrations 30, 50 and 70ng/µl with OPE 3 primer.

Table 4.18: Band Scoring Table for isolates at concentrations 30, 50 and 70 ng/μl with OPE 3 primer.

Rf value	Band (bp)	Concentration 30 ng/μl			Concentration 50ng/μl			Concentration 70ng/μl		
		1. R1	2. PS1	3. TS1	4. R1	5. PS1	6. TS1	7. R1	8. PS1	9. TS1
0.2068	1850		✔							
0.2496	1300		✔							
0.2563	1230	✔								
0.2650	1145			✔						
0.3824	435		✔							
	Total number of bands	1	3	1						
	Clarity	1	2	2						

From figure 4.15 and table 4.18, it is clear that bands were not visualised at concentrations 50 and 70 ng/μl with OPE 3 primer. At concentration 30ng/ μl, bands were visualised with each isolates. At this particular concentration, there is a high sensitivity (i.e. the presence of bands that are relatively small in size) as a band of size 435bp from PS1 isolate was present. At DNA concentration of 70ng/μl, there was the presence of smears that could be from unamplified residues.

4.4.4 Isolates R1, PS1, TS1 at concentrations 30, 50 and 70ng/µl with OPE 4 primer

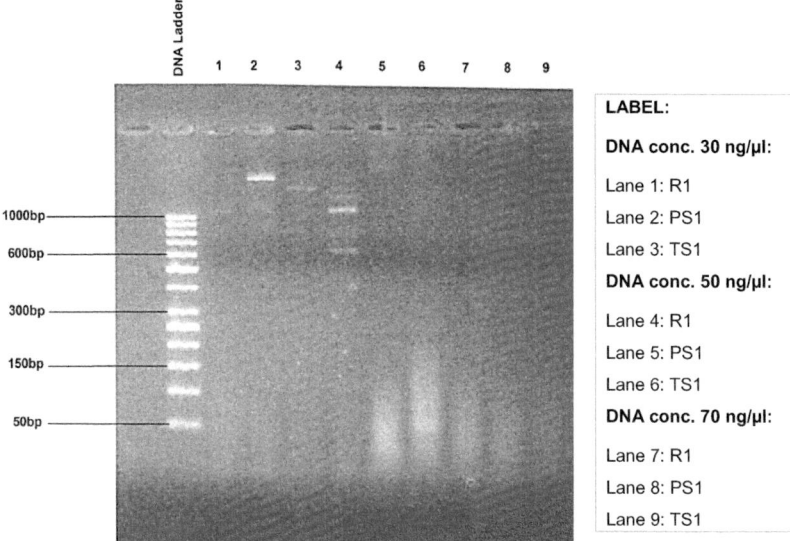

Figure 4.17: RAPD fingerprints of isolates R1, PS1 and TS1 at concentrations 30, 50 and 70ng/µl with OPE 4 primer.

Table 4.19: Rf values for the corresponding bands of the 50bp DNA Ladder

Band (DNA Ladder)	Rf value	Base Pairs
1	0.233	1000
2	0.246	900
3	0.261	800
4	0.278	700
5	0.298	600
6	0.324	500
7	0.357	400
8	0.397	300
9	0.423	250
10	0.455	200
11	0.493	150
12	0.536	100
13	0.594	50

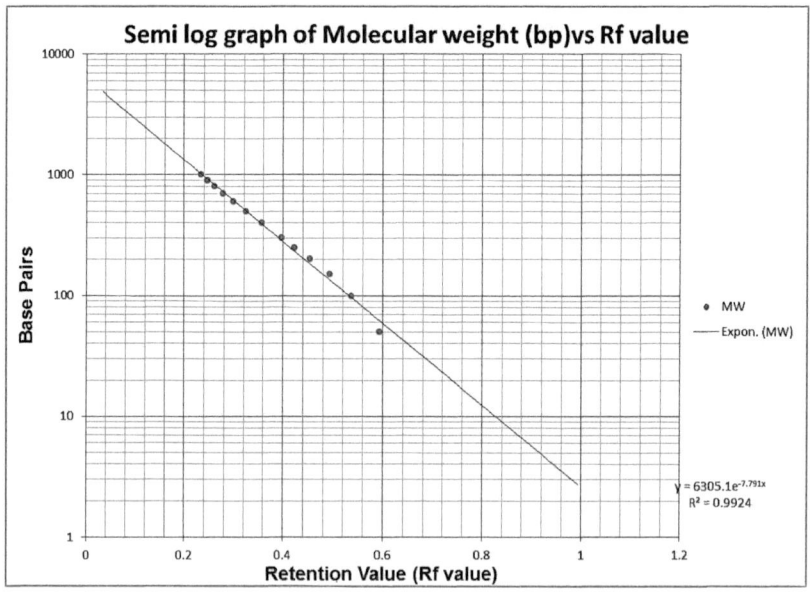

Figure 4.18: Semi log graph of Molecular weight (bp) vs Rf value for the estimation of the molecular weight of the DNA bands obtained at DNA concentrations 30, 50 and 70ng/µl with OPE 4 primer.

Table 4.20: Band Scoring Table for isolates at concentrations 30, 50 and 70 ng/µl with OPE 4 primer.

Rf value	Band (bp)	Concentration 30 ng/µl			Concentration 50ng/µl			Concentration 70ng/µl		
		1. R1	2. PS1	3. TS1	4. R1	5. PS1	6. TS1	7. R1	8. PS1	9. TS1
0.1649	1745		✔							
0.1817	1530			✔						
0.1954	1375				✔					
0.2200	1135			✔						
0.2223	1115	✔			✔					
0.2916	650				✔					
	Total number of bands	1	2	1	3					
	Clarity	3	1	2	2					

Firstly, from figure 4.17 and table 4.20, it is seen that bands were present for every isolates only at a DNA concentration of 30ng/µl. At DNA concentration of 50ng/µl, bands were visualized only with isolate R1. At DNA concentration of 30ng/µl, the clarity ranged from 1 to 3, with a clarity of 1 obtained with isolate PS1 and a clarity value of 3 being assigned to isolate R1 at concentration 30ng/µl.

4.4.5 Isolates R1, PS1, TS1 at concentrations 30, 50 and 70ng/µl with OPL 2 primer

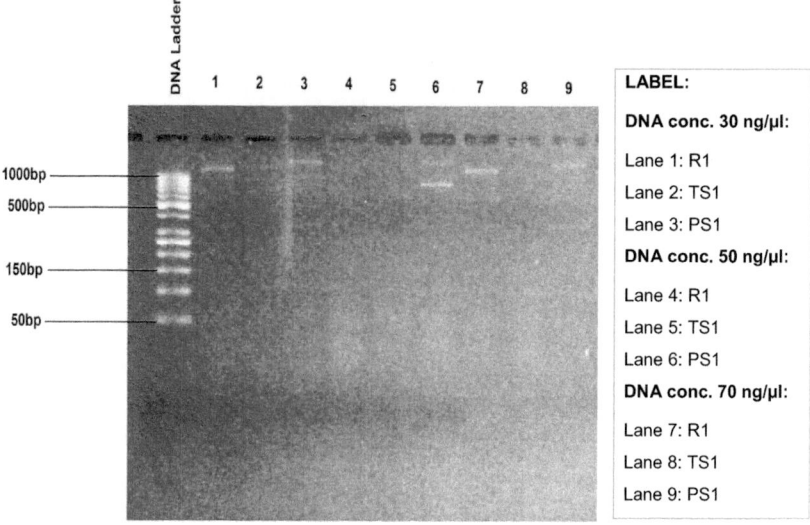

Figure 4.19: RAPD fingerprints of isolates R1, PS1 and TS1 at concentrations 30, 50 and 70ng/µl with OPL 2 primer.

Table 4.21: Rf values for the corresponding bands of the 50bp DNA Ladder.

Band (DNA Ladder)	Rf value	Base Pairs
1	0.198	1000
2	0.210	900
3	0.219	800
4	0.228	700
5	0.245	600
6	0.266	500
7	0.287	400
8	0.324	300
9	0.349	250
10	0.377	200
11	0.413	150
12	0.460	100
13	0.530	50

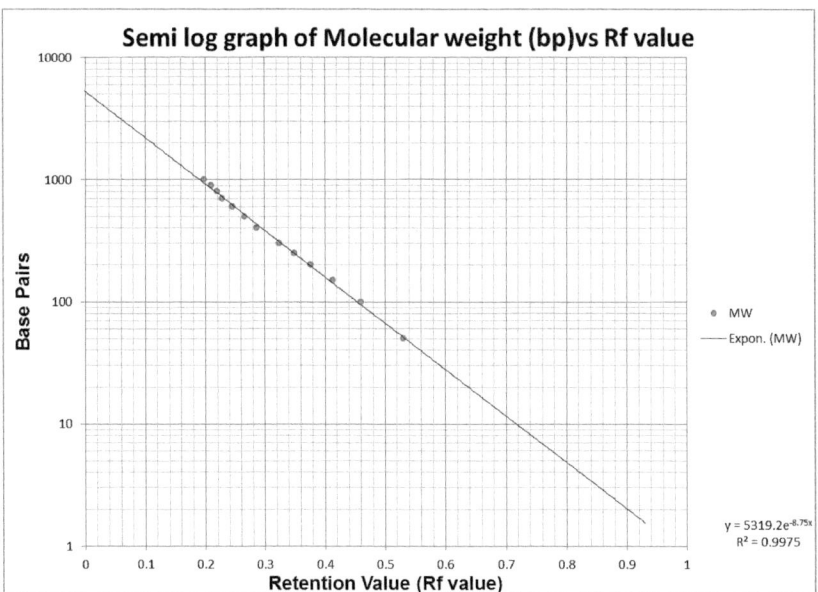

Figure 4.20: Semi log graph of Molecular weight (bp) vs Rf value for the estimation of the molecular weight of the DNA bands obtained at DNA concentrations 30, 50 and 70ng/μl with OPL 2 primer.

Table 4.22: Band Scoring Table for isolates at concentrations 30, 50 and 70 ng/µl with OPL 2 primer.

Rf value	Band (bp)	Concentration 30 ng/µl			Concentration 50ng/µl			Concentration 70ng/µl		
		1. R1	2. TS1	3. PS1	4. R1	5. TS1	6. PS1	7. R1	8. TS1	9. PS1
0.1646	1260			✔			✔			
0.1730	1170									✔
0.1832	1070	✔						✔		
0.2150	810						✔			
0.3077	360									✔
0.3109	350	✔								
	Total number of bands	2		1			2	1		2
	Clarity	1		1			1	1		2

As seen in figure 4.19 and table 4.22, bands were formed in a scattered manner without any particular trend. At DNA concentration of 30ng/µl, bands were visualised only with isolate R1 and PS1 and both were assigned a clarity value of 1. At DNA concentration of 50ng/µl, bands were formed only with isolate PS1 and it was assigned a value of 1. At DNA concentration of 70ng/µl, bands were visualised with isolate R1 and PS1. Both were assigned a clarity value of 1 and 2 respectively. It is important to note that a DNA concentration of 30ng/µl, has a better sensitivity due to the presence of a band with low size (350bp).

4.4.6 Isolates R1, PS1, TS1 at concentrations 30, 50 and 70ng/µl with OPL 4 primer

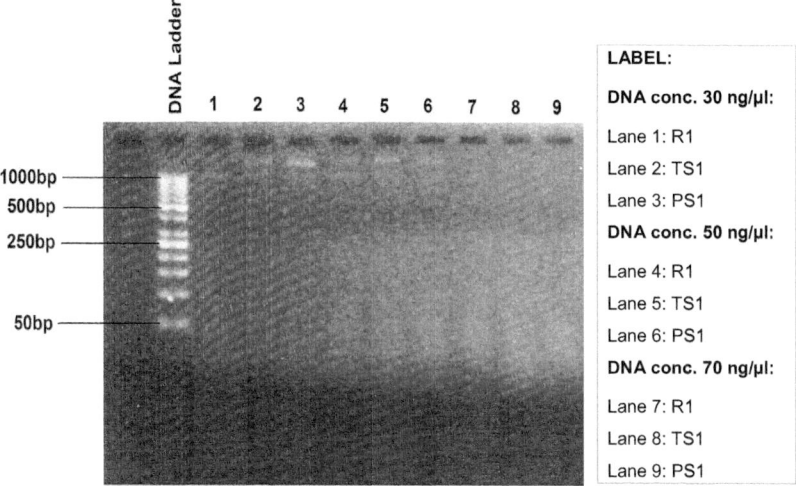

Figure 4.21: RAPD fingerprints of isolates R1, PS1 and TS1 at concentrations 30, 50 and 70ng/µl with OPL 4 primer.

Table 4.23: Rf values for the corresponding bands of the 50bp DNA Ladder.

Band (DNA Ladder)	Rf value	Base Pairs
1	0.159	1000
2	0.167	900
3	0.178	800
4	0.192	700
5	0.212	600
6	0.232	500
7	0.265	400
8	0.307	300
9	0.339	250
10	0.374	200
11	0.416	150
12	0.476	100
13	0.557	50

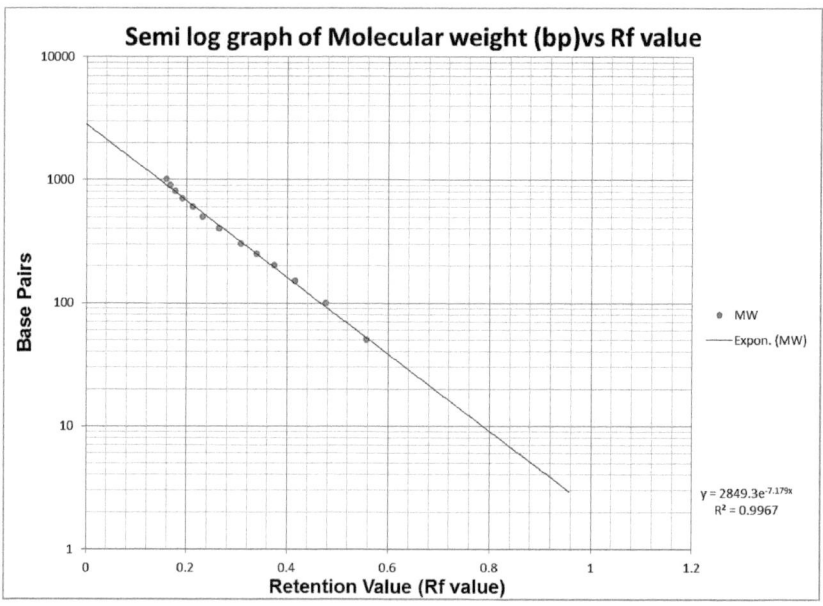

Figure 4.22: Semi log graph of Molecular weight (bp) vs Rf value for the estimation of the molecular weight of the DNA bands obtained at DNA concentrations 30, 50 and 70ng/μl with OPL 4 primer.

61

Table 4.24: Band Scoring Table for isolates at concentrations 30, 50 and 70 ng/µl with OPL 4 primer.

Rf value	Band (bp)	Concentration 30 ng/µl			Concentration 50ng/µl			Concentration 70ng/µl		
		1. R1	2. TS1	3. PS1	4. R1	5. TS1	6. PS1	7. R1	8. TS1	9. PS1
0.1087	1305		✔		✔	✔				
0.1125	1270			✔			✔			
0.1403	1040		✔							
0.1458	1000	✔			✔					
0.1501	970			✔			✔			
	Total number of bands	1	2	2	2	1	2			
	Clarity	3	2	1	2	1	2			

From figure 4.21 and table 4.24, it is clear that bands were visualised only at DNA concentrations 30 and 50ng/µl. The size of the bands ranged from 970 to 1305bp. Smears were present at DNA concentration of 70ng/µl. The clarity for isolate PS1 was assigned a value of 1 at concentration 30ng/µl. Isolates R1 and TS1 were assigned a clarity value of 3 and 2 respectively at concentration 30ng/µl. At concentration 50ng/µl, the isolates R1 and PS1 were assigned a clarity value of 2 and the isolate TS1 was assigned a clarity value of 1.

4.4.7 Synthesis of result obtained at DNA concentrations 30, 50 and 70ng/µl

In order to get a better overall understanding of the results obtained, a table has been made where all the data have been summarised concisely. The clarity (i.e. the appearance of bands in a lane relative to the background) is also evaluated. In fact, in assessing which DNA template concentration gives the best visual interpretation, the clarity can serve as a good and rapid measure.

Table 4.25: The table shows a summary of the results obtained during the testing of DNA concentrations of 30, 50 and 70 ng/µl. The clarity and frequency of clarity were evaluated.

Clarity	Frequency of clarity obtained for each isolate at different DNA concentrations								
	Concentration 30ng/µl			Concentration 50ng/µl			Concentration 70ng/µl		
	R1	PS1	TS1	R1	PS1	TS1	R1	PS1	TS1
1 (Best)	2	3	1	2	2	3	1	1	-
2 (Good)	2	3	3	2	1	-	1	1	-
3 (Acceptable)	2	-	-	-	1	-	1	1	2

From the table, we see that overall at a DNA concentration of 30ng/µl, a clarity value of 1 was recorded 6 times out of 18 PCR reactions carried out. At DNA concentration of 50ng/µl, a clarity of 1 was recorded on 7 occasions (out of 18 PCR reactions). A clarity value of 1 at DNA concentration of 70ng/µl was only recorded 2 times.

At a DNA concentration of 30ng/µl, a clarity value of 2 was assigned 8 times compared to a DNA concentration of 50ng/µl where a clarity value of 2 was assigned only on 3 occasions. At DNA concentration of 70ng/µl, a clarity value of 2 was scored only 2 times.

A clarity value of 3 was assigned 2 times for the DNA concentration of 30ng/µl (for isolate R1). The isolate PS1 at concentration 50ng/µl was assigned a clarity value of 3 on 1 occasion. At DNA concentration 70ng/µl, a clarity value of 3 was recorded on overall 4 times.

From the data we can see that isolate TS1 was scored for clarity the least number of times compared to isolate R1 and PS1.

The data from the table was then transformed into a 3-dimensional comparative bar chart for a visual interpretation and understanding of the data.

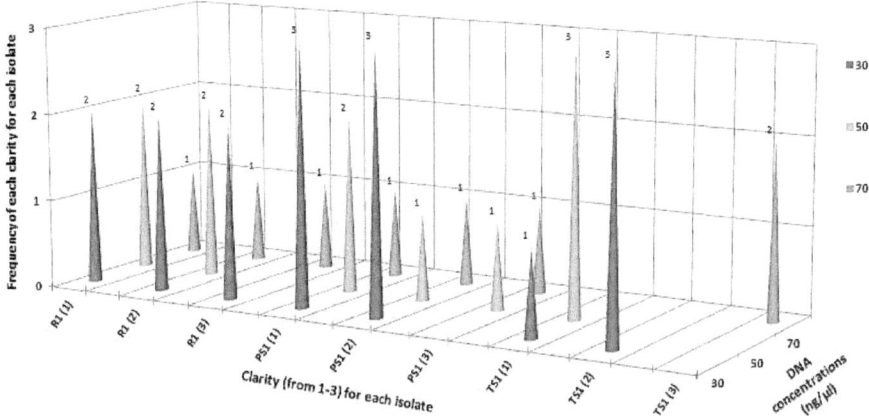

Figure 4.23: Graph of the clarity of each isolate and their frequency of occurrence during testing of DNA concentrations 30, 50 and 70ng/µl. On the x-axis, there is the clarity (from 1-3) for each isolate. On the y-axis, there is the frequency of each occurring clarity for each of the isolate and on the z-axis, there is the different DNA concentrations (30, 50 and 70ng/µl).

From the graph, we see that a DNA template concentration of 30ng/µl produced on overall, a better result with regards to the frequencies of clarity obtained for each isolates. For each of the 3 isolates at DNA concentration 30ng/µl, clarity values of 1 and 2 had been assigned. For the DNA concentration of 30ng/µl a higher number of frequencies count for clarity was recorded than for DNA concentration of 50ng/µl. The DNA concentration of 70ng/µl produced results that were the least significant as the frequencies for clarity were relatively low.

At this stage of the work, a ranking list was established where the DNA concentrations that had been tested were placed in order starting with the best DNA concentration that had been obtained so far:

1. DNA concentration of 30ng/µl
2. DNA concentration of 50ng/µl
3. DNA concentration of 70ng/µl

4.5 Testing of DNA Template Concentrations (Part 2)

For the 2nd part of the work, the DNA concentration of 40ng/μl was tested. Since DNA concentrations of 30 and 50ng/μl produced better results so far, it is necessary to check the results that could be obtained with a DNA concentration of 40ng/μl.

Concentration 40ng/μl:

1. Isolate R1: with primers OPB 5, OPB 7, OPE 3, OPE 4, OPL 2 and OPL 4
2. Isolate PS1: with primers OPB 5, OPB 7, OPE 3, OPE 4, OPL 2 and OPL 4
3. Isolate TS1: with primers OPB 5, OPB 7, OPE 3, OPE 4, OPL 2 and OPL 4

4.5.1 Isolate R1 at DNA concentration 40ng/μl

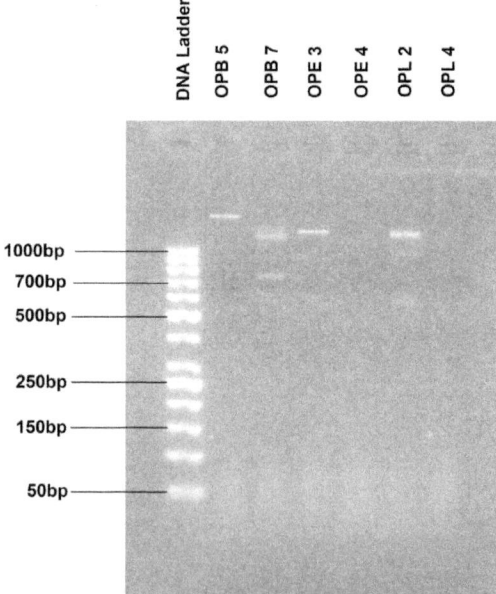

Figure 4.24: Gel image of the isolate R1 with primers OPB 5, OPB 7, OPE3, OPE 4, OPL 2 and OPL 4 at DNA concentration of 40ng/μl.

Table 4.26: Rf values for the corresponding bands of the 50bp DNA Ladder

Band (DNA Ladder)	Rf value	Base Pairs
1	0.274	1000
2	0.295	900
3	0.313	800
4	0.342	700
5	0.371	600
6	0.406	500
7	0.453	400
8	0.513	300
9	0.550	250
10	0.591	200
11	0.645	150
12	0.701	100
13	0.780	50

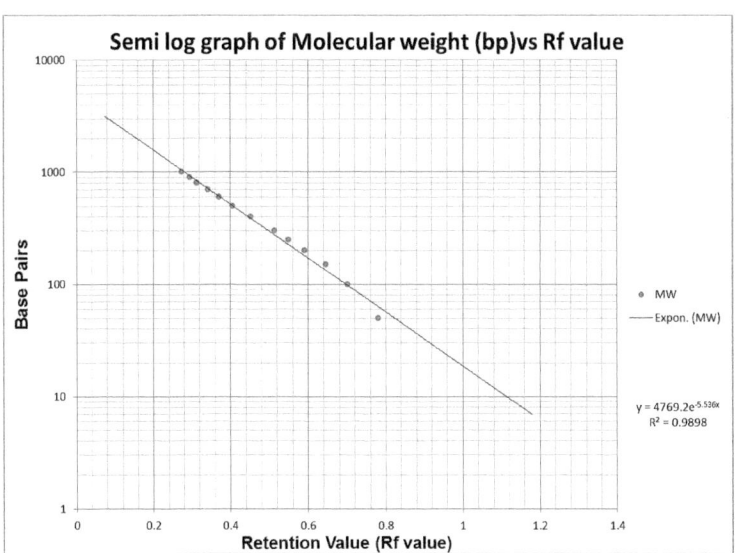

Figure 4.25: Semi log graph of Molecular weight (bp) vs Rf value for the estimation of the molecular weight of the DNA bands obtained at DNA concentration of 40ng/µl for isolate R1.

Table 4.27: Band Scoring Table for isolate R1 at concentration of 40ng/µl with the six primers.

Rf value	Band (bp)	Isolate R1 Concentration 40 ng/µl					
		OPB 5	OPB 7	OPE 3	OPE 4	OPL 2	OPL 4
0.2013	1565	✔					
0.2266	1360		✔				
0.2279	1350			✔			
0.2354	1295					✔	
0.2432	1240		✔				
0.2785	1020					✔	
0.3225	800		✔				
	Total number of bands	1	3	1		2	
	Clarity	1	1	1		1	

From figure 4.24 and table 4.27, we see that amplified fragments varied in size from 800 to 1565bp. Primers OPE 4 and OPL 2 did not produce any amplified fragments. A clarity value of 1 was assigned to all the lanes.

4.5.2 Isolate PS1 at DNA concentration 40ng/µl

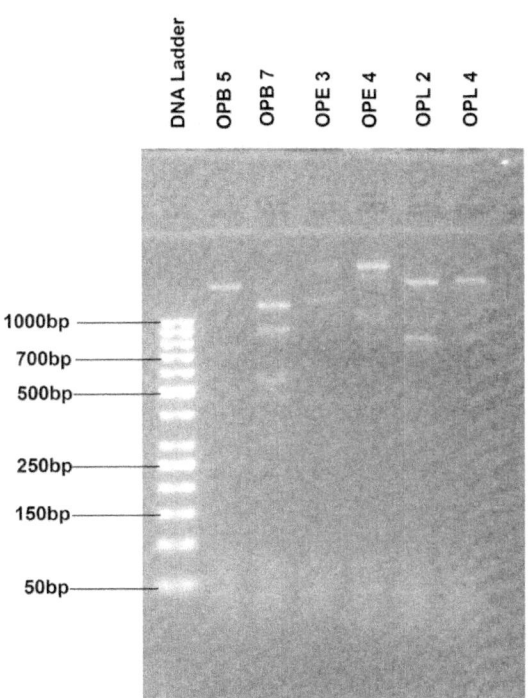

Figure 4.26: Gel image of the isolate PS1 with primers OPB 5, OPB 7, OPE3, OPE 4, OPL 2 and OPL 4 at DNA concentration of 40ng/µl.

Table 4. 28: Rf values for the corresponding bands of the 50bp DNA Ladder

Band (DNA Ladder)	Rf value	Base Pairs
1	0.273	1000
2	0.296	900
3	0.317	800
4	0.339	700
5	0.370	600
6	0.407	500
7	0.454	400
8	0.514	300
9	0.547	250
10	0.591	200
11	0.643	150
12	0.703	100
13	0.775	50

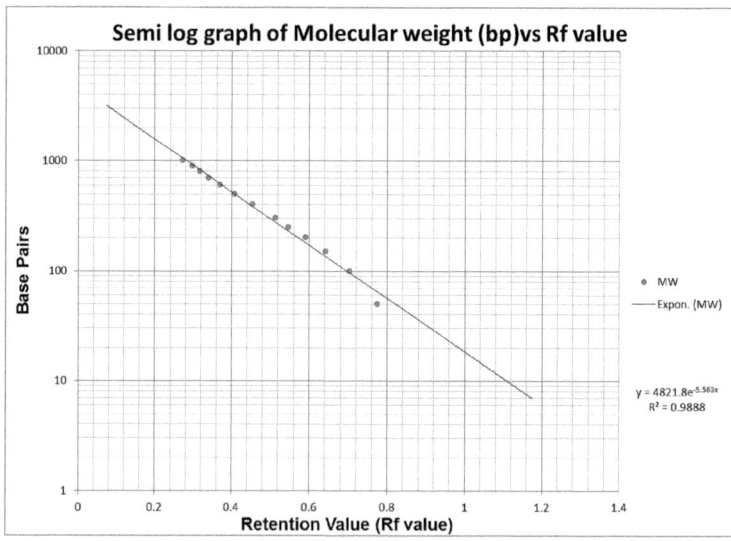

Figure 4.27: Semi log graph of Molecular weight (bp) vs Rf value for the estimation of the molecular weight of the DNA bands obtained at DNA concentration of 40ng/µl for isolate PS1.

Table 4.29: Band Scoring Table for isolate PS1 at concentration of 40ng/µl with the six primers.

Rf value	Band (bp)	Isolate PS1 Concentration 40 ng/µl					
		OPB 5	OPB 7	OPE 3	OPE 4	OPL 2	OPL 4
0.1674	1900			✔	✔		
0.1900	1675				✔		
0.1982	1600						✔
0.1999	1585					✔	
0.2099	1500	✔					
0.2349	1305			✔			
0.2426	1250		✔				
0.2929	945		✔		✔		
0.3049	884					✔	✔
0.3838	570		✔				
0.4203	465			✔			
0.4241	455		✔				
	Total number of bands	1	4	3	3	2	2
	Clarity	1	1	3	1	1	2

From figure 4.26 and table 4.29, we see that amplified fragments varied in size from 465 to 1900bp. All the primers produced positive results. The clarity values assigned were 1 (4times), 2 (1 time) and 3 (1time).

4.5.3 Isolate TS1 at DNA concentration 40ng/µl

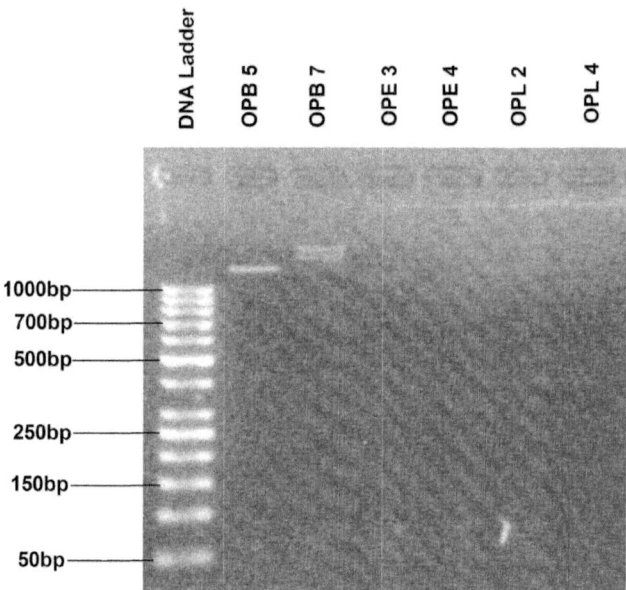

Figure 4.28: Gel image of the isolate TS1 with primers OPB 5, OPB 7, OPE3, OPE 4, OPL 2 and OPL 4 at DNA concentration of 40ng/µl.

Table 4. 30: Rf values for the corresponding bands of the 50bp DNA Ladder.

Band (DNA Ladder)	Rf value	Base Pairs
1	0.402	1000
2	0.419	900
3	0.436	800
4	0.459	700
5	0.487	600
6	0.519	500
7	0.558	400
8	0.610	300
9	0.643	250
10	0.684	200
11	0.728	150
12	0.781	100
13	0.856	50

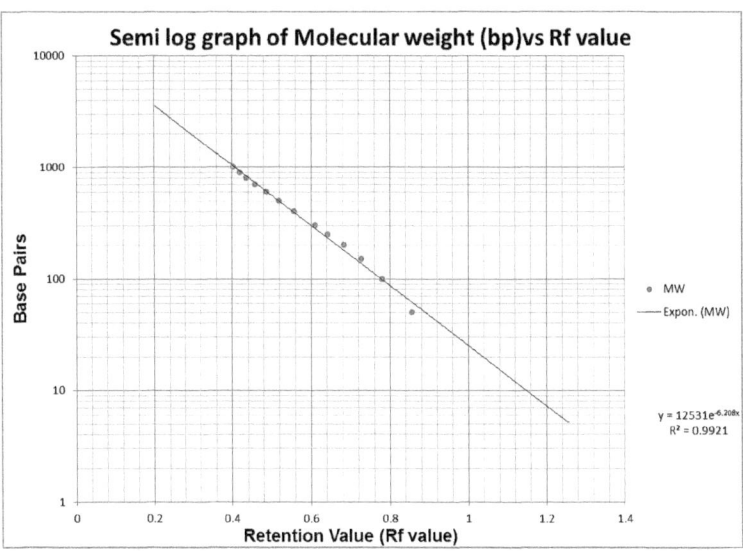

Figure 4.29: Semi log graph of Molecular weight (bp) vs Rf value for the estimation of the molecular weight of the DNA bands obtained at DNA concentration of 40ng/µl for isolate TS1.

Table 4. 31: Band Scoring Table for isolate TS1 at concentration of 40ng/μl with the six primers.

Rf value	Band (bp)	Isolate TS1 Concentration 40 ng/μl					
		OPB 5	OPB 7	OPE 3	OPE 4	OPL 2	OPL 4
0.3392	1525		✔				
0.3507	1420		✔				
0.3662	1290	✔					
	Total number of bands	1	2				
	Clarity	1	1				

From figure 4.28 and table 4.31, we see that only primers OPB 5 and OPB 7 gave positive results. The clarity values assigned were 1.

4.5.4 Synthesis of result obtained at DNA concentrations 30, 40 and 50ng/µl

Table 4.32: The table shows a summary of the results obtained during the testing of DNA concentrations of 30, 40 and 50 ng/µl. The clarity and frequency of clarity were evaluated.

Clarity	Frequency of clarity obtained for each isolate at different DNA concentrations								
	Concentration 30ng/µl			Concentration 40ng/µl			Concentration 50ng/µl		
	R1	PS1	TS1	R1	PS1	TS1	R1	PS1	TS1
1 (Best)	2	3	1	4	4	2	2	2	3
2 (Good)	2	3	3	-	1	-	2	1	-
3 (Acceptable)	2	-	-	-	1	-	-	1	-

From the table, we see that overall at a DNA concentration of 40ng/µl, a clarity value of 1 was recorded 10 times out of 18 PCR reactions carried out, compared to DNA concentration of 30ng/µl (6 times) and 50ng/µl (7 times).

At a DNA concentration of 40ng/µl, a clarity value of 2 was assigned 1 time compared to a DNA concentration of 30ng/µl where a clarity value of 2 was assigned on 8 occasions.

A clarity value of 3 was assigned only on 1 occasion for the DNA concentration of 40ng/µl (for isolate PS1). The isolate PS1 at concentration 50ng/µl was also assigned a clarity value of 3 on 1 occasion.

From the data we can see that isolate TS1 was scored for clarity the least number of times compared to isolate R1 and PS1.

The data from the table was then transformed into a 3-dimensional comparative bar chart for a visual interpretation and understanding of the data.

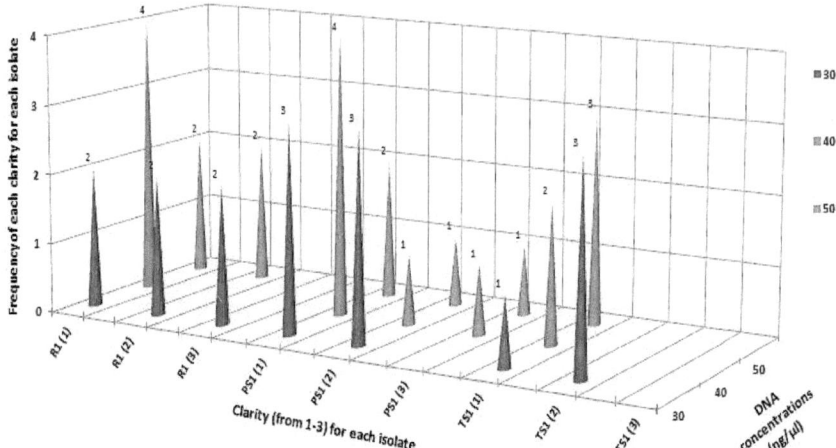

Figure 4.30: Graph of the clarity of each isolate and their frequency of occurrence during testing of DNA concentrations 30, 40 and 50ng/µl. On the x-axis, there is the clarity (from 1-3) for each isolate. On the y-axis, there is the frequency of each occurring clarity for each of the isolate and on the z-axis, there is the different DNA concentrations (30, 40 and 50ng/µl).

From the graph (figure 4.30) we see that at DNA concentration of 40ng/µl the clarity value of 1 occurs at a higher frequency (10 times for the 3 isolates combined). From the graph we also see that while a DNA concentration of 30ng/µl produced consistent results, with a concentration of 40ng/µl however, a better clarity of results (clarity value of 1) is obtained.

A ranking list was established where the DNA concentrations that had been tested were placed in order starting with the DNA concentration producing the best clarity when tested:

1.   DNA concentration of 40ng/µl

2.   DNA concentration of 30ng/µl

3.   DNA concentration of 50ng/µl

75

Chapter 5
Discussion

5.1 Analysis of Results from DNA extraction

The use of good quality DNA is crucial when doing RAPD-PCR because it is an extremely sensitive process. A carefully chosen method of genomic DNA extraction which contains a minimal amount of contaminants (proteins, ethanol, Na^+) may produce better amplification during PCR (Ramella et al., 2005). A relatively good purity of DNA (Table 4.2) and a high DNA concentration was obtained for each of the 3 isolates (Table 4.3). The purity of DNA obtained was around 1.81 to 1.95 and the DNA concentrations varied from 1025 to 5595ng/µl. This is due to the genomic DNA extraction method that has been used. Grinding with mortar and pestle is an effective way of breaking the fungal cell walls (Karakousis et al., 2006). The tandem usage of ethanol and isopropanol for precipitation has also helped in obtaining a high amount of pure DNA material. A high concentration of DNA is obtained when using cold isopropanol for precipitation and washing with 70% ethanol causes the removal of salts which are potential PCR inhibitors (Cenis, 1992; Aljanabi & Martinez, 1997). For the long term of storage DNA, TE buffer was preferred over distilled water because it maintains the stability of the DNA (Oxford Gene Technology, 2011).

5.2 Primers used

Single arbitrary decamers are used to amplify RAPD marker fragments (Mitchelson et al., 1999). During PCR, the random oligonucleotide will bind to specific sequences on each genomic DNA strands in opposite orientations and within an amplifiable distance which is generally less than 1500bp (Kumar & Gurusubramanian, 2011; Semagn et al., 2006).

Screening of the primers was made at a DNA concentration of 50ng/µl which was chosen at random. High concentrations of DNA such as 100 and 200ng/µl produced smears only and a low DNA concentration of 20ng/µl did not produce any amplification of marker fragments. From the screening step, only 6 primers produced RAPD marker fragments, OPB 5 and OPB 7 had a G-C content of 70% and OPE 3, OPE 4, OPL 2 and OPL 4 had a G-C content of 60%. Knowledge of the % G-C content of a primer is important because G-C pairing consists of 3 hydrogen bonds compared to A-T pairing that contains only 2 hydrogen bonds and stronger hydrogen bridges may enable the template DNA-primer-DNA polymerase complex to resist a temperature of 72°C during the elongation phase (William et al., 1990). These 6 primers were used for testing of different DNA template concentrations.

5.3 Analysis of the Results from testing of DNA concentrations

After the screening step, the DNA concentration of 50ng/μl served as a reference concentration and further testing of DNA concentrations have been based around this value. For instance, it is of good practice to test a two-fold dilution and a two-fold concentration around that reference DNA concentration (i.e. 50ng/μl) (Saunders & Hopkins, 1999, p.120). However during the screening step it has been seen that an approximate two-fold concentration (80 and 100ng/μl) and an approximate two-fold dilution (20ng/μl) did not yield any amplified products. High DNA template concentrations only produced smears whereas low DNA template concentrations did not undergo any amplification. This is because PCR is an enzymatic based reaction whereby the interaction of the primer, DNA template and DNA polymerase is sensitive to changes in each of these parameters.

RAPD-PCR being a laboratory dependent assay, it was necessary to devise a setup where different DNA concentrations could be compared simultaneously after PCR and to enable scoring. The DNA concentrations 30, 40, 50 and 70ng/μl produced various degrees of amplifications. In this work, the amplified fragments of each lane as a whole were assessed for clarity and were taken into consideration instead of assessing individual amplification fragments.

5.3.1 DNA template concentration of 70ng/μl

According to Fraga *et al.* (2005), for a wide range of organisms, between 10 and 100ng of DNA template with high purity is sufficient to obtain high quality and reproducible RAPD profile. Consequently, in this present optimization protocol, a DNA template concentration of 70ng/μl (70ng/μl of DNA x 5.0μl per reaction mix) was considered relatively too high and out of a true amplifiable DNA concentration range, because it lacked reproducibility and a frequency of clarity value of 3 ('acceptable' level) was more commonly recorded at this concentration. In some cases amplification did not occur at all or fragments greater than 1000bp were more commonly obtained. Also, to be noted is that out of the six trials at DNA concentration 70ng/μl, and except for the OPL 2 primer (60% G-C content), amplification also occurred when using OPB 5 and OPB 7 primers which have a high (70%) G-C content and these will anneal more frequently to the DNA template strand at a wider range of DNA template concentration due to the high number of hydrogen bonds formation and there is a risk of a low primer specificity, nonspecific annealing of off target sequences (Ashrafi & Paul, 2009) and giving false-positive signal (Dobosy *et al.*, 2011) . In this particular case, this combination of DNA concentration of 70ng/μl with 70% G-C content primer has been found to be unsuitable for RAPD-PCR of *P.infestans* because the conditions may not be suitable for successful amplification. This high DNA concentration also disrupts amplification because of competition occurring for the primer to bind to the template DNA (Micheli *et al.*, 1994).

77

5.3.2 DNA template concentrations of 30, 40 and 50ng/µl

In this project, DNA template concentrations of 30, 40 and 50ng/µl were considered as part of an acceptable range of DNA concentrations that gave strong results with the RAPD-PCR assay. From table 4.32 and figure 4.30, we see that that at concentrations 30 and 50ng/µl, a clarity value of 1 (best visual appearance) was obtained 6 and 7 times respectively whereas at concentration 40ng/µl a clarity value of 1 was obtained 10 times (out of 18 PCR reactions at each DNA concentrations). Actually at concentration 40ng/µl a clarity value of 1 was most commonly obtained whereas at concentrations of 30 and 50ng/µl, clarity values of 1, 2 and 3 were homogeneously distributed as shown in table 4.32. Hence, the optimal DNA template concentration in this laboratory protocol was 40ng/µl and concentrations 30 and 50ng/µl served as extremes in that range. However it should also be noted that at DNA concentrations 30 and 40ng/µl, low molecular weight fragments (300-600bp) were obtained especially with OPB 7, OPE 3 and OPL 2 primers.

A study of 141 isolates of *Phytophthora infestans* using RAPD in Canada between 1994 and 1996 made use of an optimal 20ng of DNA of *P.infestans* in 25µl master mix volume (Mahuku *et al.*, 2000). However in that particular study, a more refined protocol was used; the DNA extraction was carried out with DNAzol® Reagent (Life Technologies) (Mahuku *et al.*, 1998) and the composition of the master mix was fully optimized. In another study by Samen *et al.* (2003), only 25ng of genomic DNA was used in 12.5 µl master mix volume. In contrast, in this project work, no DNA purification was made and a range of 150ng (30ng/µl of DNA x 5.0µl per reaction mix) up to 250ng (50ng/µl of DNA x 5.0µl per reaction mix) was used to generate RAPD fingerprints. This particular comparison between this project and the work of Fraga *et al.* (2005), Mahuku *et al.* (2000) and Samen *et al.* (2003) focuses on two particular important points: (1) the amount of genomic DNA of *Phytophthora infestans* required for RAPD-PCR depends on the extraction protocol used and the presence of any inhibitors, for example, the amount of genomic DNA used in RAPD-PCR may be relatively less when purification is made using DNAzol® Reagent (Life Technologies) than with other methods and (2) the optimization of all the parameters and consequently the use of a standard protocol may lower the optimal DNA template concentration required for amplification during PCR. In this project work, a DNA concentration of 40ng/µl, is deemed optimal, however as previous literatures suggest, optimization of all the parameters in RAPD may lower that optimal DNA concentration which is required for the reproducibility of RAPD. A lower concentration of DNA in a fully optimized protocol may be a potential advantage when having to deal with certain isolates that generate only a small amount of DNA after genomic DNA extraction.

5.4 Other potential key factors affecting the RAPD assay

For a RAPD protocol to be considered fully reproducible, the results obtained in one laboratory must be obtained as it is in other laboratories (Penner *et al.*, 1993). In this present work, a combination of parameters was used from different studies such that a specific protocol may be produced for the study of the genetic diversity of *Phytophthora infestans* on the island. Parameters like the concentration of $MgCl_2$ (1.5mM from 5X Phusion Green HF Buffer), concentration of *Taq* DNA polymerase (0.2U) were as suggested by Hussain *et al.* (2014) and concentration of dNTPs (0.2mM) was as used by Mahuku *et al.* (2000) but unlike in these previous studies mentioned above, a relatively higher RAPD primer concentration of 2.0µM was used mainly due to the fact that this protocol was not yet optimized and that this particular constant concentration of primer of 2.0µM in fact maximized the possibility of obtaining amplification products while doing the testing of different DNA template concentrations. However in this project work, there are several other key parameters apart from DNA template concentrations that require a particular attention and which may affect amplification of genomic DNA.

5.4.1 The annealing temperature

A 35°C annealing temperature was used in this protocol. This particular temperature is in the range of the melting points of the primers used (Promega, no date). A much higher annealing temperature prevents the amplification by the primers. A low annealing temperature was used to favour a maximum of primer-DNA template interaction and the generation of amplified fragments (Savva *et al.*, 2000). The number of thermal cycle is also of key importance. In this study, a total of 32 cycles were used to get identifiable amplified fragments. A higher number of cycles (>40) usually result in the formation of nonspecific fragments (Roche Life Science, no date).

5.4.2 TE buffer for storage of DNA and its effect on Mg^{2+} availability

For the storage of DNA, a pH of 8.0 is used as nucleases are less active at this pH and the EDTA in TE buffer reduces base hydrolysis by chelating divalent metal ions such as Mg^{2+} which are required for nucleases activity (GeneLink, no date). The dilution of DNA extract from historic specimens of *Phytophthora infestans* in TE buffer for PCR normally required 10 mM of Tris-HCl, pH 8.0 and 0.1 mM of EDTA (Ristaino *et al.*, 2001). However, in this project work, 1 mM of EDTA was used as suggested by Goodwin *et al.* (1992) and by El Komy *et al.* (2012) for DNA extraction from fresh specimens and long term storage. More than 1mM of EDTA is not advisable to use for PCR (MRC Holland, no date). On the other hand, the presence of a relatively high amount of Mg^{2+} causes nonspecific annealing of primers (Roche Life Science, no date). In this work, EDTA may have potentially interacted with and chelated the Mg^{2+} present in the PCR mix. MgCl$_2$ acts as a co-factor of the *Taq* polymerase and a lack of Mg^{2+} may have impacted on the amplification of DNA templates (Ramella *et al.*, 2005).

5.5 Isolates used in this study and Genetic diversity

From this study, we noted a difference in RAPD fingerprints between R1 and PS1 that were isolated from potato (*Solanum tuberosum L.*) and TS1 which was isolated from tomato (*Solanum lycopersicum*). The selection of the primers was made based on their screening with isolate R1. However when testing the different DNA template concentrations of the isolates, there was evidence that not all the selected primers took part in DNA amplification with the TS1 isolate. Even at the optimal concentration of 40ng/μl only 2 primers (OPB 5 and OPB 7) produced amplified fragments with the TS1 isolate. However as stated by Yildirim *et al.* (2007), the use of only 6 RAPD primers is not enough to conclude that there is indeed genetic variation occurring among the isolates affecting tomato and potato. Therefore, optimization of all parameters involved in the RAPD-PCR protocol, as previously mentioned, is required before any claims can be made regarding genetic variations among the 3 tested isolates in this study.

Chapter 6
Conclusion &
Recommendations

6.1 Conclusion

The development of the RAPD (random amplified polymorphic DNA) method enabled the scoring of organisms at numerous loci and hence genetic diversity studies (Lynch & Milligan, 1994). RAPD markers are suitable for the study of asexually reproducing organisms such as *Phytophthora infestans* (Bardakci, 2001). This method is cheap, fast and performed much easily than other methods used for genetic studies (Bardakci, 2001).

For the RAPD method to be fully functional and reliable, conditions like DNA template concentration, primer concentration, thermocycler conditions, Taq polymerase concentration and Mg^{2+} concentration should be at optimal level.

The aim of this project was to determine which template DNA concentration of *Phytophthora infestans* gives the best reproducible results with RAPD-PCR in order to develop a reproducible protocol for the genetic diversity studies of local populations of *P.infestans*. Relatively high concentrations of DNA such as 70, 80, 100 and 200ng/µl did not commonly produced amplification. It was however found that the amplification of DNA in RAPD worked well at a concentration ranging from 30 to 50ng/µl. At a DNA concentration of 30ng/µl consistent RAPD fingerprints were obtained but it was found that at a DNA concentration of 40ng/µl a better clarity of the amplified fragments in the lanes was obtained. Hence, the DNA concentration of 40ng/µl is considered as being optimal in this protocol.

The use of RAPD fingerprinting in Mauritius can prove to be very useful due to its low cost and fast accessibility to results. Since it is probable that only the A1 mating type of *Phytophthora infestans* is present on the island (Global Initiative Late Blight, 2004), the asexually reproducing oomycete may be investigated using the RAPD method which can detect polymorphic fragments among the isolates around the island and hence determine clonal identity (Bardakci, 2001). By combining the genetic profiles of the isolates available around the island to their susceptibility to fungicide applications a more effective management of the late blight disease can be made.

6.2 Recommendations

Considering the results of this study, several recommendations can be made in order to acquire more feasible results with respect to the optimization of the DNA template concentration in RAPD for the genetic characterization of *P. infestans*.

1. Instead of using TE buffer for the storage of DNA, the disaccharide trehalose may be used for preserving the DNA quality and not to interfere during PCR (Smith & Morin, 2005). Trehalose can also enhance PCR and provide thermal stability to Taq polymerase (Spiess *et al.*, 2004).

2. More primers and isolates could be used in the assessment of the optimal concentration of DNA. This would in turn increase the pool of data obtained for clarity and allow a more confident selection of the optimal concentration of DNA.

3. In order to increase the potential of the PCR reaction, three primers (triple RAPD PCR) can be used and a combination of new bands may be formed (Mansour *et al.*, 2008). Here also, a higher number of data may help in discriminating among the concentrations that may seem to be optimal.

4. For an assessment of the genetic diversity of *Phytophthora infestans* on the island, a good practice would be to use a combination of different molecular markers such as RAPD (random amplified polymorphic DNA) and ISSR (inter-simple sequence repeat) techniques to confirm the level of genetic variation (Yildirim *et al.*, 2007) as they are simple to generate.

References

List of References:

1. AGRIOS, G., 2005. *Plant Pathology*. 5th ed. San Diego: Academic Press.

2. ALJANABI, S. M., AND MARTINEZ, I., 1997. Universal and rapid salt-extraction of high quality genomic DNA for PCR-based techniques. *Nucleic acids research*, 25(22), 4692-4693.

3. ALLERS, T., AND LICHTEN, M., 2000. A method for preparing genomic DNA that restrains branch migration of Holliday junctions. *Nucleic acids research*, 28(2), e6.

4. ASHRAFI, E. H., AND PAUL, N., 2009. Improved PCR specificity with Hot Start PCR primers. *BioTechniques*, 47(3), 789–790.

5. ATHEYA, I., SINGH, B. P., CHAKRABARTI, S. K., AND PATTANAYAK, D., 2005. Genetic diversity and differentiation of Indian isolates of *Phytophthora infestans* as revealed by RAPD analysis. *Indian journal of experimental biology*, 43(9), 817.

6. AYLOR, D. E., FRY, W. E., MAYTON, H., AND ANDRADE-PIEDRA, J. L., 2001. Quantifying the rate of release and escape of *Phytophthora infestans* sporangia from a potato canopy. *Phytopathology*, 91(12), 1189-1196.

7. BARDAKCI, F., 2001. Random Amplified Polymorphic DNA (RAPD) Markers. *Turkish Journal of Biology*, 25, 185-196.

8. BIRCH, P., AND WHISSON, S., 2001. *Phytophthora infestans* enters the genomics era. *Molecular Plant Pathology*, 2(5), 257-263.

9. BROOKES, A. J., 1999. The essence of SNPs. *Gene*, 234(2), 177-186.

10. BURGES, N., TAYLOR, A., MACKIE, A., AND KUMAR, S., 2005. *Phytophthora infestans*-Exotic threat to Western Australia Factsheet. Department of Agriculture, The State of Western Australia.

11. CARRIS, L. M., LITTLE, C. R., AND STILES, C. M., 2012. Introduction To Fungi. *Plant Health Instructor, The American Phytopathological Society* [online].
Available from:
http://www.apsnet.org/edcenter/intropp/PathogenGroups/Pages/IntroFungi.aspx
[Accessed 30 Nov. 2014]

12. CENIS, J. L., 1992. Rapid extraction of fungal DNA for PCR amplification. *Nucleic acids research*, 20(9), 2380.

13.　CÉSPEDES, M. C., CÁRDENAS, M. E., VARGAS, A. M., ROJAS, A., MORALES, J. G., JIMÉNEZ, P., BERNAL, A. J., AND RESTREPO, S., 2013. Physiological and molecular characterization of *Phytophthora infestans* isolates from the Central Colombian Andean Region. *Revista Iberoamericana de Micología*, 30(2), 81-87.

14.　COOKE, D.E.L., AND LEES, A.K., 2004. Markers, old and new, for examining *Phytophthora infestans* diversity. *Plant Pathology*, 53, 692-704.

15.　DIAS-NETO, E., DE SOUZA, C. P., ROLLINSON, D., KATZ, N., PENA, S. D., AND SIMPSON, A. J. , 1993. The random amplification of polymorphic DNA allows the identification of strains and species of schistosome. *Molecular and Biochemical Parasitology*, 57(1), 83-88.

16.　DOBOSY, J., ROSE, S., BELTZ, K., RUPP, S., POWERS, K., BEHLKE, M., AND WALDER, J., 2011. RNase H-dependent PCR (rhPCR): improved specificity and single nucleotide polymorphism detection using blocked cleavable primers. *BMC Biotechnology*, 11(1), 80. doi:10.1186/1472-6750-11-80

17.　DRENTH, A., AND SENDALL, B., 2001. Practical guide to detection and identification of *Phytophthora*. *Brisbane (Australia)*. CRA for Tropical Plant Protection. 41p.

18.　EDWARDS, K. J., AND MOGG, R., 2001. Plant Genotyping by Analysis of Single Nucleotide Polymorphisms. *In:*　R. J. HENRY, ed. *Plant genotyping: the DNA fingerprinting of plants,* London: CABI publishing, 1-3.
Available from:
https://books.google.mu/books?hl=en&lr=&id=OyymA1RBhbEC&oi=fnd&pg=PR10&d q=Plant+genotyping:+the+DNA+fingerprinting+of+plants&ots=GPI6XTLcQC&sig=_7jn wUR0wmZeK7ciOEYKDROv_Y0&redir_esc=y#v=onepage&q=Plant%20genotyping% 3A%20the%20DNA%20fingerprinting%20of%20plants&f=false　[Accessed 15 Feb. 2015]

19.　EJAZ, M., GAISHENG, Z., NA, N., HUIYAN, Z., QIDI, Z., AND QUNZHU, W., 2014. Comparison of small scale methods for the rapid and efficient extraction of mitochondrial DNA from wheat crop suitable for down-stream processes. *Genetics and molecular research: GMR*, 13(4), 10320-10331.

20.　EL-KOMY, M. H., SALEH, A. A., AND MOLAN, Y. Y., 2012. Molecular characterization of early blight disease resistant and susceptible potato cultivars using random amplified polymorphic DNA (RAPD) and simple sequence repeats (SSR) markers. *African Journal of Biotechnology*, 11(1), 37-45.

21.   FRAGA, J., RODRIGUEZ, J., FUENTES, O., FERNANDEZ-CALIENES, A., AND CASTEX, M., 2005. Optimization of random amplified polymorphic DNA techniques for use in genetic studies of Cuban Triatominae. *Revista do Instituto de Medicina Tropical de São Paulo*, 47(5), 295-300.

22.   FRY, W. E., GOODWIN, S.B., DYER, A.T., MATUSZAK, J.M., DRENTH, A., TOOLEY, P.W., SUJKOWSKI, L.S., KOH, Y.J., COHEN, B.A., SPIELMAN, L.J., DEAHL, K.L., INGLIS, D.A. AND SANDLAN, K.P., 1993. Historical and recent migrations of *Phytophthora infestans*: Chronology, pathways and implications. *Plant Disease*, 77, 653-661.

23.   FRY, W. E., AND SMART, C. D., 1999. The return of *Phytophthora infestans*, a potato pathogen that just won't quit. *European Association of Potato Research* [online], 42, 279-282.
       Available from:
       http://link.springer.com/article/10.1007/BF02357858#page-1 [Accessed 30 Oct. 2014]

24.   FRY, W. E., SHAW, D., AND FLIER, D., 2007. Laboratory Manual for *P. infestans* work at CIP (Centro Internacional de la Papa) [online].
       Available from:
       https://research.cip.cgiar.org/typo3/web/fileadmin/icmtoolbox/ICM_Toolbox/Files/Manual_draft1.pdf [Accessed 30 Nov. 2014].

25.   GENELINK, no date. *TE Buffer 1X Solution pH 8.0; 50 ml* [online].
       Available from:
       http://www.genelink.com/geneprodsite/product.asp?p=11675 [Accessed 27 Feb. 2015].

26.   GLOBAL INITIATIVE LATE BLIGHT, 2004. *Mauritius Late Blight Profile* [online].
       Available from:
       https://research.cip.cgiar.org/confluence/display/GILBWEB/Mauritius [Accessed 18 Dec. 2014]

27.   GOODWIN, S. B., DRENTH, A., AND FRY, W. E., 1992. Cloning and genetic analyses of two highly polymorphic, moderately repetitive nuclear DNAs from *Phytophthora infestans. Current genetics* [online], 22(2), 107-115.
       Available from:
       http://link.springer.com/article/10.1007/BF00351469#page-1 [Accessed 30 Jan. 2015]

28. GOODWIN, S., COHEN, B., AND FRY, W., 1994. Panglobal distribution of a single clonal lineage of the Irish potato famine fungus. *Proceedings of the National Academy of Sciences*, 91(24), 11591-11595.

29. GOSS, E. M., TABIMA, J. F., COOKE, D. E. L., RESTREPO, S., FRY, W. E., FORBES, G. A., FIELAND, V. J., CARDENAS, M., AND GRUNWALD, N. J., 2014. The Irish potato famine pathogen *Phytophthora infestans* originated in central Mexico rather than the Andes. *Proceedings of the National Academy of Sciences*, 111(24), 8791-8796.

30. GRÜNWALD, J. N., AND FLIER, G. W., 2005. The Biology of *Phytophthora infestans* at its Center of Origin. *Annual Review of Phytopathology*. 43, 171-190.

31. GUO, J., VAN DER LEE, T., QU, D. Y., YAO, Y. Q., GONG, X. F., LIANG, D. L., XIE, K. Y., WANG, X. W., AND GOVERS, F., 2009. *Phytophthora infestans* isolates from Northern China show high virulence diversity but low genotypic diversity. *Plant Biology*, 11(1), 57-67.

32. GUPTA, A., GUPTA, N., AND GUPTA, D.K., 2010. Molecular investigation in wistar rats by RAPD typing. *Science & Culture*, 76 (7-8), 261-266.

33. HAAS, B. J., KAMOUN, S., ZODY, M. C., JIANG, R. H., HANDSAKER, R. E., CANO, L. M., GRABHERR, M., KODIRA, C. D., RAFFAELE, S., TORTO-ALALIBO, T., BOZKURT, T. O., AH-FONG, A. M. V., ALVARADO, L., ANDERSON, V. L., ARMSTRONG, M. R., AVROVA, A., BAXTER, L., BEYNON, J., BOEVINK, P. C., BOLLMANN, S. R., BOS, J. I. B., BULONE, V., CAI, G., CAKIR, C., CARRINGTON, J. C., CHAWNER, M., CONTI, L., COSTANZO, S., EWAN, R., FAHLGREN, N., FISCHBACH, M. A., FUGELSTAD, J., GILROY, E. M., GNERRE, S., GREEN, P. J., GRENVILLE-BRIGGS, L. J., GRIFFITH, J., GRÜNWALD, N. J., HORN, K., HORNER, N. R., HU, C., HUITEMA, E., JEONG, D., JONES, A. M. E., JONES, J. D. G., JONES, R. W., KARLSSON, E. K., KUNJETI, S. G., LAMOUR, K., LIU, Z., MA, L., MAC LEAN, D., CHIBUCOS, M. C., MCDONALD, MCWALTERS, J., MEIJER, H. J. G., MORGAN, W., MORRIS, P. F., MUNRO, C. A., O'NEILL, K., OSPINA-GIRALDO, M., PINZÓN, A., PRITCHARD, L., RAMSAHOYE, B., REN, Q., RESTREPO, S., ROY, S., SADANANDOM, A., SAVIDOR, A., SCHORNACK, S., SCHWARTZ, D. C., SCHUMANN, U. D., SCHWESSINGER, B., SEYER, L., SHARPE, T., SILVAR, C., SONG, J., STUDHOLME, D. J., SYKES, S., THINES, M., VAN DE VONDERVOORT, P. J. I., PHUNTUMART, V., WAWRA, S., WEIDE, R., WIN, J., YOUNG, C., ZHOU, S., FRY, W., MEYERS, B. C., VAN WEST, P., RISTAINO, J., GOVERS, F., BIRCH, P. R. J., WHISSON, S., JUDELSON, H. S., AND NUSBAUM, C., 2009. Genome sequence

and analysis of the Irish potato famine pathogen *Phytophthora infestans. Nature*, 461(7262), 393-398.

34. HIMEDIALABS, 2011. Rye Agar B Technical data [online].
     Available from:
     http://www.himedialabs.com/TD/M1855.pdf [Accessed 16 Feb. 2015].

35. HUSSAIN, T., SINGH, B. P., AND ANWAR, F., 2014. A quantitative Real Time PCR based method for the detection of *Phytophthora infestans* causing Late blight of potato, in infested soil. *Saudi journal of biological sciences*, 21(4), 380-386.

36. ILARIONOVA, E. V., 2006. *Molecular Genetic and Functional Characterization of candidate loci for controlling quantitative resistance to the oomycete Phytophthora infestans*. Thesis (PhD), Universität zu Köln.

37. KAMOUN, S., 1998. Resistance of *Nicotiana benthamiana* to *Phytophthora infestans* is mediated by the Recognition of the Elicitor Protein INF1. *The Plant Cell Online*, 10(9), pp.1413-1426.

38. KAMOUN, S., 2003. Molecular genetics of pathogenic oomycetes. *Eukaryotic cell*, 2(2), 191-199.

39. KARAKOUSIS, A., TAN, L., ELLIS, D., ALEXIOU, H., & WORMALD, P. J., 2006. An assessment of the efficiency of fungal DNA extraction methods for maximizing the detection of medically important fungi using PCR. *Journal of microbiological methods*, 65(1), 38-48.

40. KROON, L. P., BROUWER, H., DE COCK, A. W., AND GOVERS, F., 2012. The genus *Phytophthora* anno 2012. *Phytopathology*, 102(4), 348-364.

41. KUMAR, P., GUPTA, V. K., MISRA, A. K., MODI, D. R., AND PANDEY, B. K., 2009. Potential of molecular markers in plant biotechnology. *Plant Omics J*, 2(4), 141-162.

42. KUMAR, N. S., AND GURUSUBRAMANIAN, G., 2011. Random amplified polymorphic DNA (RAPD) markers and its applications. *Science Vision*, 11(3), 116-124.

43. KUPFER, D. M., REECE, C. A., CLIFTON, S. W., ROE, B. A., AND PRADE, R. A., 1997. Multicellular ascomycetous fungal genomes contain more than 8000 genes. *Fungal genetics and biology*, 21(3), 364-372.

44. LYNCH, M., AND MILLIGAN, B. G., 1994. Analysis of population genetic structure with RAPD markers. *Molecular ecology*, 3(2), 91-99.

45. MAHUKU, G. S., HSIANG, T., AND YANG, L., 1998. Genetic diversity of *Microdochium nivale* isolates from turfgrass. *Mycological Research*, 102(05), 559-567.

46. MAHUKU, G., PETERS, R. D., PLATT, H. W., AND DAAYF, F., 2000. Random amplified polymorphic DNA (RAPD) analysis of *Phytophthora infestans* isolates collected in Canada during 1994 to 1996. *Plant Pathology*, 49(2), 252-260.

47. MANSOUR, A., ISMAIL, O. M., AND EL-DIN, S. M. M., 2008. Diversity assessments among Mango (*Mangifera indica L.*) cultivars in Egypt using ISSR and three-primer based RAPD fingerprints. *African Journal of Plant Science and Biotechnology*, 2(2), 87-92.

48. MARTIN, F. N., ABAD, Z. G., BALCI, Y., AND IVORS, K., 2012. Identification and detection of *Phytophthora*: reviewing our progress, identifying our needs. *Plant Disease*, 96(8), 1080-1103.

49. MAURITIUS SUGAR INDUSTRY RESEARCH INSTITUTE, 2000. Annual Report. Reduit, Mauritius.

50. MAZÁKOVÁ, J., TABORSKY, V., ZOUHAR, M., RYSANEK, P., HAUSVATER, E., AND DOLEZAL, P., 2006. Occurrence and distribution of mating types A1 and A2 of *Phytophthora infestans* (Mont.) de Bary in the Czech Republic. *Plant Protection Science*. 42(2), 41-48.

51. METAPATHOGEN, no date, *Phytophthora infestans, late blight, potato blight: facts, life cycle, mating types, tissues at MetaPathogen* [online].
     Available from:
     http://www.metapathogen.com/phytophthora/ [Accessed 18 Mar. 2015].

52. MICHELI, M. R., BOVA, R., PASCALE, E., AND D'AMBROSIO, E., 1994. Reproducible DNA fingerprinting with the random amplified polymorphic DNA (RAPD) method. *Nucleic Acids Research*, 22, 1921-1922.

53. MITCHELSON K.R., DRENTH J., DUONG H., CHAPARRO J.X., 1999. Direct sequencing of RAPD fragments using 3'-extended oligonucleotide primers and dye terminator cycle-sequencing. *Nucleic acids research*, 27(19), pp.28e-28.

54. MOLECULARCLONING, 2012. Precipitation of DNA with Isopropanol [online].
     Available from:
     http://www.molecularcloning.com/index.php?prt=5 [Accessed 20 Mar. 2015].

55.   MRC HOLLAND, 2009. *Sample treatment* [online].
       Available from:
       http://www.mlpa.com/WebForms/WebFormMain.aspx?Tag=_wl2zCji-
       rCGANQgZPuTixn01pTTvcNt-SNVkcQD5y3xyfzzlPululw.. [Accessed 26 Feb. 2015].

56.   OPERON TECHNOLOGIES INC., no date. *Operon and BC primers sequences.*
       California, USA [online].
       Available from:
       http://www.vcru.wisc.edu/simonlab/sdata/pdf/RAPDs-
       OperonandBCPrimerSequences.xls [Accessed 20 Mar. 2015].

57.   OXFORD GENE TECHNOLOGY, 2011. *DNA Storage And Quality* [online].
       Available from:
       http://www.ogt.co.uk/resources/literature/403_dna_storage_and_quality. [Accessed 26
       Feb. 2015].

58.   PENNER, G., BUSH, A., WISE, R., KIM, W., DOMIER, L., KASHA, K., LAROCHE, A.,
       SCOLES, G., MOLNAR, S. AND FEDAK, G., 1993. Reproducibility of random
       amplified polymorphic DNA (RAPD) analysis among laboratories. *Genome Research*,
       2(4), 341-345.

59.   PROMEGA, no date, *How do I determine the concentration, yield and purity of a DNA
       sample?* [online].
       Available from:
       http://worldwide.promega.com/resources/pubhub/enotes/how-do-i-determine-the-
       concentration-yield-and-purity-of-a-dna-sample/ [Accessed 17 Feb. 2015].

60.   PROMEGA, no date, *BioMath - Tm Calculations for Oligos* [online].
       Available from:
       http://www.promega.com/a/apps/biomath/index.html?calc=tm [Accessed 27 Feb.
       2015].

61.   QIAGEN, no date. Why do I have to add beta-mercaptoethanol (beta-ME) to lysis
       Buffer RLT of the RNeasy Kits? [online].
       Available from:
       http://www.qiagen.com/us/resources/faq?id=eac3139e-6c6c-4172-b61f-
       18d61b6cdd1e [Accessed 16 Feb. 2015].

62.   RAMELLA, M. S., KROTH, M. A., TAGLIARI, C., AND ARISI, A. C. M., 2005.
       Optimization of random amplified polymorphic DNA protocol for molecular
       identification of *Lophius gastrophysus*. *Food Science and Technology (Campinas)*,
       25(4), 733-735.

63. RISTAINO, J. B., GROVES, C. T., AND PARRA, G. R., 2001. PCR amplification of the Irish potato famine pathogen from historic specimens. *Nature*, 411(6838), 695-697.

64. RISTAINO, J. B., 2002. Tracking historic migrations of the Irish potato famine pathogen, *Phytophthora infestans*. *Microbes and Infection*, 4(13), 1369-1377.

65. ROCHE LIFE SCIENCE, no date. *Working with PCR* [online].
Available from:
http://lifescience.roche.com/shop/en/us/products/working-with-pcr [Accessed 26 Mar. 2015].

66. ROYCHOWDHURY, R., TAOUTAOU, A., HAKEEM, K. R., GAWWAD, M. R. A., AND TAH, J., 2013. Molecular Marker-Assisted Technologies for Crop Improvement. In *Crop Improvement in the Era of Climate Change*.

67. SAMEN, A. E., SECOR, G. A., AND GUDMESTAD, N. C., 2003. Genetic variation among asexual progeny of *Phytophthora infestans* detected with RAPD and AFLP markers. *Plant Pathology*, 52(3), 314-325.

68. SAUNDERS, G.C, AND HOPKINS D., 1999. Random Amplified Polymorphic DNA (RAPD) Analysis. In: Saunders, G.C., and Parkes, H.C. eds. *Analytical Molecular Biology: Quality and Validation*. Royal Society of Chemistry, pp: 103-120.
Available from:
https://books.google.mu/books?id=_yU4J0vl0WgC&pg=PR1&dq=Analytical+Molecular+Biology:+Quality+and+Validation&hl=en&sa=X&ei=HRIUVYGhJsb2UK6gg7AH&ved=0CBsQ6AEwAA#v=onepage&q=Analytical%20Molecular%20Biology%3A%20Quality%20and%20Validation&f=false [Accessed 30 Nov. 2014]

69. SAVVA, D., DEPLEDGE, M., ATIENZAR, F., EVENDEN, A., AND JHA, A., 2000. Optimized RAPD analysis generates high-quality genomic DNA profiles at high annealing temperature. *Biotechniques*, 28(1), 52-54.

70. SCHUMANN, G.L., AND D'ARCY, G., 2000. *Late blight of potato and tomato. The Plant Health Instructor*. The American Phytopathological Society [online].
Available from:
https://www.apsnet.org/edcenter/intropp/lessons/fungi/Oomycetes/Pages/LateBlight.aspx [Accessed 16 Oct. 2014].

71. SEMAGN, K., BJØRNSTAD, Å., AND NDJIONDJOP, M. N., 2006. An overview of molecular marker methods for plants. *African Journal of Biotechnology*, 5(25).

72. SHAW D., AND WATTIER R., 2003. Evolution of *Phytophthora infestans*: a global overview. *In*: C. LIZARRAGA, ed. *Proceedings of the Global Initiative on Late-Blight (GILB) Conference 11–13 July, 2002, Hamburg, Germany*. Lima, Peru: CIP, 23–27. Available from: https://books.google.mu/books?id=kWkkSyQP82IC&pg=PR3&dq=Proceedings+of+the+Global+Initiative+on+Late-Blight+%28GILB%29+Conference&hl=en&sa=X&ei=uBUUVe2yH4G0ULaNgWA&ved=0CBsQ6AEwAA#v=onepage&q=Proceedings%20of%20the%20Global%20Initiative%20on%20Late-Blight%20(GILB)%20Conference&f=false [Accessed 30 Jan. 2015]

73. SHIHAB, K.M., AND AHMAD S., 2014. Eco-Friendly Management Of Late Blight Of Tomato *(Lycopersicon esculentum L.)*. *International Journal of Agricultural Science and Research (IJASR)*, 4(3), 165-174.

74. SIMON FRASER UNIVERSITY, no date. *BISC 429: Experimental Techniques II Separation Methods* [online]. Available from: http://www.sfu.ca/bisc/bisc-429/DNA1.html#intro [Accessed 20 Mar. 2015].

75. SKORIĆ, M., ŠILER, B., BANJANAC, T., ŽIVKOVIĆ, J. N., DMITROVIĆ, S., MIŠIĆ, D., AND GRUBIŠIĆ, D., 2012. The reproducibility of RAPD profiles: effects of PCR components on RAPD analysis of four *Centaurium* species. *Archives of Biological Sciences*, 64(1), 191-199.

76. SMITH, S., AND MORIN, P. A., 2005. Optimal storage conditions for highly dilute DNA samples: a role for trehalose as a preserving agent. *Journal of forensic sciences*, 50(5), 1101-1108.

77. SOGIN, M. L., AND SILBERMAN, J. D., 1998. Evolution of the protists and protistan parasites from the perspective of molecular systematics. *International journal for parasitology* [online], 28(1), 11-20. Available from: http://www.sciencedirect.com/science/article/pii/S0020751997001811 [Accessed 30 Sept. 2014]

78. SONI, N.K., AND SONI, V., 2010. Fundamentals of Botany. New Delhi, Tata McGraw-Hill.
Available from:
https://books.google.mu/books?id=RpwLnPQEPgYC&pg=PT4&lpg=PT4&dq=Fundam entals+of+Botany.+New+Delhi,+Tata+McGraw-+Hill.&source=bl&ots=7mt_87ceR0&sig=h57Ub0Brm1uxSv-e8YPFflYEkXU&hl=en&sa=X&ei=ehgUVZ6yEoqrUYy7g9gK&ved=0CCAQ6AEwAQ#v =onepage&q=Fundamentals%20of%20Botany.%20New%20Delhi%2C%20Tata%20M cGraw-%20Hill.&f=false [Accessed 14 Jan. 2015]

79. SPIESS, A. N., MUELLER, N., AND IVELL, R., 2004. Trehalose is a potent PCR enhancer: lowering of DNA melting temperature and thermal stabilization of Taq polymerase by the disaccharide trehalose. Clinical Chemistry, 50(7), 1256-1259.

80. THE PENNSYLVANIA STATE UNIVERSITY, 1998. Identifying Potato Diseases in Pennsylvania. Pennsylvania: The Pennsylvania State University [online].
Available from:
http://pubs.cas.psu.edu/FreePubs/pdfs/agrs75.pdf pdf [Accessed 30 Nov. 2014]

81. THERMO SCIENTIFIC, no date. GeneRuler 1 kb DNA Ladder, ready-to-use 250 to 10,000 bp [online].
Available from:
http://www.thermoscientificbio.com/nucleic-acid-electrophoresis /generuler-1-kb-dna-ladder-ready-to-use-250-to-10000-bp/ [Accessed 29 Dec. 2014].

82. THERMO SCIENTIFIC, no date. GeneRuler 50 bp DNA Ladder, ready-to-use, 50-1000 bp [online].
Available from:
http://www.thermoscientificbio.com/nucleic-acid-electrophoresis /generuler-50-bp-dna-ladder-ready-to-use/ [Accessed 30 Dec. 2014].

83. VOLK, J. T., 2001. Phytophthora infestans, cause of late blight of potato and Irish potato famine. University of Wisconsin, La Crosse [online].
Available from:
http://botit.botany.wisc.edu/toms_fungi/m2001alt.html [Accessed 30 Oct. 2014].

84. WHITE, T., ADAMS, W., AND NEALE, D., 2007. *Forest genetics*: CABI, Wallingford, UK: 66-67.

Available from:

https://books.google.mu/books?id=UHZCeg4BqtkC&pg=PR4&dq=Forest+genetics:+C ABI,+Wallingford,+UK:&hl=en&sa=X&ei=oRwUVdzFGsHsUNyugpgN&ved=0CD8Q6A EwBw#v=onepage&q=Forest%20genetics%3A%20CABI%2C%20Wallingford%2C%2 0UK%3A&f=false [Accessed 18 Nov. 2014]

85. WILLIAMS, J. G., KUBELIK, A. R., LIVAK, K. J., RAFALSKI, J. A., AND TINGEY, S. V., 1990. DNA polymorphisms amplified by arbitrary primers are useful as genetic markers. *Nucleic acids research*, 18(22), 6531-6535.

86. XIAO QIONG, Z., YINGHUA, W., AND LIYUN, G., 2006. Genetic diversity revealed by RAPD analysis among isolates of *Phytophthora infestans* from different locations in China. *Acta Phytopathologica Sinica* [online]. 36(3), 249-258.

Available from:

http://www.cabdirect.org/abstracts/20063137848.html [Accessed 3 Feb. 2015]

87. YILDIRIM, N. T. A., TURKUSAY, H., AND TANYOLAC, B., 2007. Genetic variation among *Phytophthora infestans* (tomato blight) isolates from western Turkey revealed by inter simple sequence repeat (ISSR) and random amplified polymorphic DNA (RAPD) markers. *Pakistan Journal of Botany*, 39(3), 897-902.

Appendix

Appendix 3.3

Table A 1: The properties of the antibiotics used

Antibiotic	Activity	Target organisms
Vancomycin	Antibacterial	Gram+ve bacteria; Gram-ve bacteria to a lesser extent
Polymixin B	Antibacterial	Gram –ve bacteria
Ampicillin	Antibacterial	Gram +ve bacteria
Rifampicin	Antibacterial	Gram +ve bacteria; Gram –ve bacteria to a lesser extent
Pentachloronitrobenzene (PCNB)	Antifungal	Narrow antifungal spectrum
Benomyl	Antifungal	Most fungi except Zygomycetes and Oomycetes

Source (Drenth & Sendall, Practical guide to detection and identification of *Phytophthora*, 2001, online)

Appendix 3.6.1

Table A 2: Composition of Extraction buffer and the concentration of stock solutions

Buffer	Chemical	Final Concentration	Stock Solution
Extraction	EDTA	0.05M	0.5M
	Tris pH 8.0	0.1M	1M
	NaCl	0.5M	5M
	Beta mercaptoethanol	0.7%	Added to final after autoclave
	SDS	0.25%	2.5%

Source: (Laboratory Manual for *P. infestans* work, International Potato Center, 2007)

EDTA:

1000mL= 0.5moles	**In 100mL Extraction Buffer**
using $C_1V_1=C_2V_2$	using $C_1V_1=C_2V_2$
Mass= number of moles x Mr	0.5M x V_1= 0.05M x 100mL
0.5M x V_1= 0.05M x 100mL	V_1 = (0.05M x 100mL)/0.5M
= 0.5moles x 372.24g	V_1= 10mL
V_1 = (0.05M x 100mL)/0.5M	
= 186.12g	
1000mL= 186.12g	
100mL= 18.612g	

Tris pH 8.0:

1000mL= 1 mole	**In 100mL Extraction Buffer**
Mass=number of moles x Mr	using $C_1V_1=C_2V_2$
=1moles x 157.6g	1M x V_1= 0.1M x 100mL
=157.6g	V_1 = (0.1M x 100mL)/1M
1000mL= 157.6g	V_1= 10mL
100mL= 15.76g	

NaCl:

1000mL= 5 moles	**In 100mL Extraction Buffer**
Mass= number of moles x Mr	using $C_1V_1=C_2V_2$
= 5 moles x 58.44g	5M x V_1= 0.5M x 100mL
= 292.2g	V_1 = (0.5M x 100mL)/5M
1000mL= 292.2g	V_1= 10mL
100ml= 29.22g	

Beta mercaptoethanol:

Add 0.7 mL in final 100mL after autoclaving

SDS:

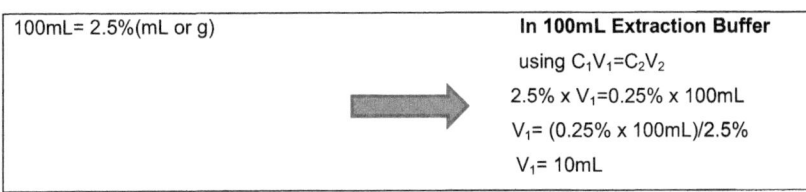

100mL= 2.5%(mL or g)	**In 100mL Extraction Buffer**
	using $C_1V_1=C_2V_2$
	2.5% x V_1=0.25% x 100mL
	V_1= (0.25% x 100mL)/2.5%
	V_1= 10mL

Table A 3: Composition of TE buffer and the concentration of stock solutions

Buffer	Chemical	Final Concentration	Stock Solution
TE	EDTA	10mM	100mM
	Tris pH 8.0	1mM	10mM

Source: (Laboratory Manual for *P. infestans* work, International Potato Center,2007)

YOUR KNOWLEDGE HAS VALUE

- We will publish your bachelor's and
  master's thesis, essays and papers

- Your own eBook and book -
  sold worldwide in all relevant shops

- Earn money with each sale

Upload your text at www.GRIN.com
and publish for free